engineering
soil
mechanics

CIVIL ENGINEERING AND ENGINEERING MECHANICS SERIES

N. M. Newmark and W. J. Hall, editors

PRENTICE-HALL INTERNATIONAL, INC. London
PRENTICE-HALL OF AUSTRALIA, PTY. LTD., Sydney
PRENTICE-HALL OF CANADA, LTD., Toronto
PRENTICE-HALL OF INDIA PRIVATE LIMITED, New Delhi
PRENTICE-HALL OF JAPAN, INC., Tokyo

engineering soil mechanics

JAN J. TUMA

Engineering Consultant
Boulder, Colorado

M. ABDEL-HADY

Professor of Civil Engineering
Oklahoma State University

PRENTICE-HALL, INC., Englewood Cliffs, New Jersey

10 9 8 7 6 5 4 3 2 1

ISBN: 0-13-279505-1

Library of Congress Catalog Card Number: 72-6460

Printed in the United States of America

TO OUR TEACHERS OF SOIL MECHANICS

contents

6

volume change

7

stress analysis

8

strength analysis

9
stability analysis

10
frost action

preface

This book presents in one volume a systematic summary of the major informations and tools of engineering soil mechanics and was prepared with the intent to serve as a textbook for a formal college course in engineering soil mechanics at the elementary and intermediate level.

The subject matter is divided into ten chapters forming three distinct groups, organized in logical sequences. The first group of chapters includes the study of morphology, methods of classification, rheology, investigation of soil water and of soil consolidation. The second group presents three types of analysis of soil systems: the stress analysis, the strength analysis, and the stability analysis. Finally, the study of the frost action is introduced in the last chapter as a separate topic.

The organization of this material and the form of presentation has many special features facilitating the learning process and allowing an easy and rapid location of desired information.

(1) Each chapter consists of three parts: the theory, the solved problems, and the supplementary problems.

(2) The theory part of each chapter introduces the definitions and classification of major terms, the statements of theorems with their analytical formulations and/or with their experimental verifications, and outlines in broad forms the methods of solution or of investigation of particular classes of problems.

(3) The material of the respective theory part is then supported by deriva-

tions, proofs, experiments, discussions and numerical examples in the solved problems part.

(4) Simultaneously, additional examples and discussions with indicated solutions and answers are given in the supplementary problems part of the same chapter.

This method of presentation allows the inclusion of a large amount of material, considerably more than can be covered in one course at a given level. Consequently, some topics such as the study of the rheological models (Chapter 3), the analysis of half-space (Chapter 7), the stress tensor concept (first part of Chapter 8), and the study of frost action (Chapter 10) may and should be omitted in the introductory course and reintroduced at the intermediate level or used as special reading assignments.

In the preparation and organization of the material presented in this book, the authors have relied on an extensive body of knowledge developed by two generations of soil mechanicists and foundation engineers. Contrary to the established practice to credit these sources by names only (such as, after Terzaghi, Casagrande, Taylor, etc.) a complete reference is given in each case of adopted material in the text and all such references are included in the final list of references at the end of this book.

Boulder, Colorado   J. J. Tuma

Stillwater, Oklahoma   M. Abdel-Hady

engineering soil mechanics

morphology

1

1.1 INTRODUCTION

Object

The *earth's crust* is made up of *rocks* and *soils*. The latter category consists of systems of solid particles derived from the rocks by the action of a series of weathering processes and includes organic matters, water, air (gases), and bacteria. The systematic investigation of the behavior of rocks and soils is called rock mechanics and soil mechanics, respectively. If this study is related to engineering applications, the designations of *engineering rock mechanics* and *engineering soil mechanics* are used (Problem 1.1).

Although rocks and soils are used for many different purposes, their main functions in engineering are to serve as the *foundation* and *construction materials*.

Analysis

The material in this book is restricted to the analysis of stability, rigidity, and strength of soils in foundations and earth structures. Numerical, graphical, and experimental methods are available for this purpose.

Since soil mechanics is only an extension and further development of the principles of applied mechanics of solids, fluids, and gases, the governing relationships used in this analysis are based on

1. The principles of static and dynamic equilibrium.
2. The principles of compatibility of dislocation.
3. The principles of conservation and dissipation of energy.
4. The constitutive laws relating causes and effects (derived from experiments).

This enumeration indicates the restrictive nature of the elementary analytical model and the consequent oversimplification of the problem at this level of study.

1.2 TYPES OF FOUNDATION MATERIAL

Function of Foundation

The *structure* (building, bridge, retaining wall, pavement slab, etc.) is a mechanism designed and built to carry loads and resist forces (caused by volume change and self–stressing). In order to perform this function, the structure must be supported by a medium (foundation) capable of absorbing this transmission (of bearing this effect). The analytical prediction of this capability (or incapability) is the ultimate aim of engineering soil mechanics.

Macroscopic Classification

The visual inspection of the construction side (of foundation material) is the first, and an absolutely necessary, step in achieving this aim. For convenience, the foundation materials are grouped into the following series of types, easily detectable by macroscopic (visual and manual) examination.

1. *Bedrock* or *ledge rock* is sound, hard, undistributed rock lying in the location of its formation and underlain by no other material but rock.
2. *Shattered* or *broken rock* is the transition-zone material between soils and bedrock, consisting of large and small fragments of rock frequently filled with seams of softer material.
3. *Boulders* are fragments or rounded pieces of rock over 10 in. in maximum dimension. *Cobbles* are 2 to 10 in. in size, and *pebbles* are 4 mm to 2 in. in size. Boulders occur singly or form whole fields accompanied by cobbles, pebbles, and sand.
4. *Gravels* are pieces of rock (2 mm to 6 in.) broken away from bedrock and forming layers or pockets. *Pea gravel* is the designation of fragments 2 mm to $\frac{1}{4}$ in. in size.
5. *Sand* consists of small rounded rock particles that vary in size from 0.05 to 2.00 mm. Frequently three subdivisions are used in the description of sand: fine sand (0.05 to 0.20 mm), medium sand (0.20 to 0.60 mm), and coarse sand (0.60 to 2.00 mm).
6. *Silt* is a system of fine particles (powder) of rock of size 0.005 to 0.050 mm, some of which cannot be readily distinguished macroscopically.

7. *Clay* is a system of exceedingly fine inorganic particles of less than 0.005 mm in size. Clays may be classified as soft, medium, or stiff, depending on the moisture content and degree of consolidation (Chapter 6). A *heavy-textured alluvial clay* is called *adobe* and *fine-textured greasy clay* is known as *gumbo*.

8. *Hardpan* is a mixture of gravel and sand with silt and/or clay cemented together in a mass of high density. *Caliche* is a lighter mixture of gravels, sands, and silts only partially cemented together by salts during the evaporation process.

9. *Loam* is a mixture of sand, silt, and/or clay with some organic matters called *humus*. This type of soil is frequently called the *topsoil*.

10. *Loess* is a wind-blown, fine-grained soil consisting of particles of less than 0.05 mm in size.

Soils that are *unsuitable as foundation materials* are *mud* (or *muck*), a sticky mixture of earth and water, *peat*, partly decayed organic matters, and *betonite*, fine-grained volcanic ash.

Graphical Representation

Although there is a great inconsistency in the graphical representation of these various soils, the symbols shown in Fig. 1-1 are believed to be those used by the majority of engineers.

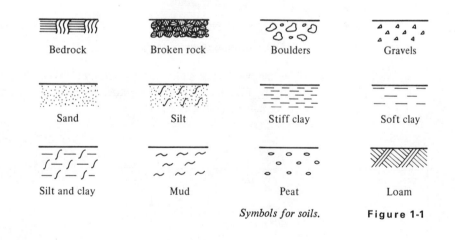

Bedrock	Broken rock	Boulders	Gravels
Sand	Silt	Stiff clay	Soft clay
Silt and clay	Mud	Peat	Loam

Symbols for soils. **Figure 1-1**

Solids in Soil

The solid particles forming the types of soil introduced above fall into two categories: *products of rock weathering* and *products of organic decay*. The weathering process is the breakdown of rocks into smaller pieces by *disintegration* (mechanical process), *decomposition* (chemical process), and *solution* (chemical and mechanical process). The organic decay is a process of *decomposition of vegetable matters* buried by soil or intermixed with it. Since this process produces acid-reducing environment, it works as a *catalyzer* in the rock–weathering process (Problems 1.2 and 1.3).

Rock and Clay Minerals

Five most important rock minerals are the *silicas*, the *feldspars*, the *micas*, the *ferromagnesians*, and the *carbonates*, among which various forms of silica are the most common soil solids (Problem 1.4).

Three most important clay minerals are the *kaolinites*, the *montmorillonites*, and the *illites*, all of which are highly complex hydroaluminum silicates (Problems 1.5 and 1.6).

1.3 BASIC PROPERTIES

Three-phase Composition

The volume V_m of a given soil sample is

$$V_m = V_s + V_v = V_s + V_w + V_g \tag{1-1}$$

where V_s = volume of solids,

V_v = volume of voids,

V_w = volume of water,

V_g = volume of gases (air).

The composition of soil volume given by (1-1) can be represented diagrammatically by Fig. 1-2, which is, of course, only an idealization (since all voids and solids cannot be segregated).

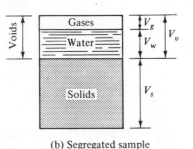

(a) True sample

(b) Segregated sample

Figure 1-2   *Three-phase soil sample.*

The weight W_m of the same soil sample is

$$W_m = W_s + W_w + W_g \cong W_s + W_w \qquad (1\text{-}2)$$

where W_s = weight of solids,

$\quad W_w$ = weight of water,

$\quad W_g = 0$ = weight of gases (air), which is disregarded as insignificant.

Volume Relationships

Three volume ratios derived from Fig. 1-2 are of practical significance—the *porosity*, the *void ratio*, and the *degree of saturation*.

The porosity n is the ratio of the volume of voids V_v to the total volume V_m.

$$n = \frac{V_v}{V_m} \qquad (1\text{-}3)$$

The void ratio e is the ratio of the volume of voids V_v to the volume of solids V_s.

$$e = \frac{V_v}{V_s} = \frac{n}{1-n} \qquad (1\text{-}4)$$

The degree of saturation S is the ratio of the volume of water V_w to the volume of voids V_v.

$$S = \frac{V_w}{V_v} \qquad (1\text{-}5)$$

In these relationships [(1-3), (1-4), (1-5)], e is usually given as a dimensionless coefficient, whereas n and S are expressed as percentages (Problems 1.7 and 1.8).

Weight Relationships

Three terms related to the weight of the soil sample are frequently used: the *unit weight*, the *specific gravity*, and the *water content*.

The unit weight γ_m of the soil sample is the weight of its unit volume; that is, it is the ratio of the total weight W_m to the total volume V_m, also called the *density*.

$$\gamma_m = \frac{W_m}{V_m} \qquad (1\text{-}6)$$

This relationship is perfectly general, and for the given sample,

$$\gamma_s = \frac{W_s}{V_s}, \quad \gamma_w = \frac{W_w}{V_w}, \quad \gamma_d = \frac{W_s}{V_m} \tag{1-7}$$

where γ_s = unit weight of solids,

γ_w = unit weight of water,

γ_d = dry unit weight (dry density).

The specific gravity G_m of the soil sample is the ratio of its weight to the weight of an equal volume of distilled water at 4 °C.

$$G_m = \frac{W_m}{V_m \gamma_0} = \frac{\gamma_m}{\gamma_0} \tag{1-8}$$

where γ_0 is the unit weight of distilled water at 4 °C.

Similarly,

$$G_s = \frac{W_s}{V_s \gamma_0} = \frac{\gamma_s}{\gamma_0} \qquad G_w = \frac{W_w}{V_w \gamma_0} = \frac{\gamma_w}{\gamma_0} \tag{1-9}$$

where G_s is the specific gravity of solids and G_w is the specific gravity of water.

The water content w of the soil sample is the ratio of the weight of water W_w to the weight of solids W_s.

$$w = \frac{W_w}{W_s} \tag{1-10}$$

In these relationships [(1-6) through (1-10)], γ is given in pounds per cubic foot (or in grams per cubic centimeter), G is a dimensionless ratio, and w is given in percentages (Problems 1.9, 1.10, 1.11). Numerically, the unit weight and the specific gravity are equal values in the metric system but they are different values in the U.S. system.

1.4 SOIL STRUCTURE

Classification

The arrangement of soil components (solids, water, air voids) is known as the *structure of soil*. Obviously this arrangement can take a limitless number of forms. The following three categories are typical: *cohesionless soil, cohesive soil*, and *skeletal soil* (Problems 1.16 and 1.17).

Cohesionless Soils

The cohesionless soils (Fig. 1-3) are composed of bulky grains loosely or densely packed together. They are frequently called single-grain soils (such as sands

(a) Loose

(b) Dense

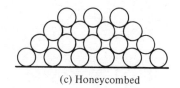

(c) Honeycombed

Cohesionless soil structures.   **Figure 1-3**

and gravels), and their stability is directly related to their density. The cohesionless soils owe their shearing resistance to the interparticle friction.

Cohesive Soils

The cohesive soils (Fig. 1-4) are composed of clay minerals that introduce a complex system of forces (attractions and repulsions) within the soils, thus

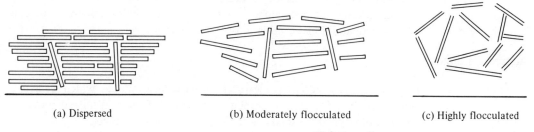

(a) Dispersed   (b) Moderately flocculated   (c) Highly flocculated

Cohesive soil structures.   **Figure 1-4**

leading to the dispersed or flocculent arrangements. The cohesive soils offer a shear resistance in the absence of the lateral pressure and are, in general, good substructures.

Skeletal Soils

The skeletal soils (Fig. 1-5) consist of a framework of large grains (gravels) held together by a binder. This composition usually occurs in one of the three typical forms: *matrix structure, contact-bound structure,* and *void-bound structure.* The strength and stability of this system are conditioned by the properties of the skeleton but primarily by the properties of the binder.

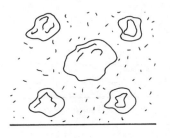
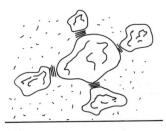

(a) Matrix structure   (b) Contact-bound structure   (c) Void-bound structure

Figure 1-5   *Skeletal soil structures.*

1.5 SOIL CONSISTENCY

Concept

The soil consistency refers to the condition of soil defined by such terms as *liquid*, *plastic*, or *solid* (*hard*). Obviously these definitions have a meaning only in cases of fine-grain soils, the moisture content of which affects the consistency. Yet the moisture content alone is not an adequate index (measure) of consistency, and additional criteria must be introduced.

Atterberg's Limits

The consistency generally is measured by Atterberg's system, which recognizes four states: liquid, plastic, semisolid, and solid. A fine-grain soil, such as clay, will pass through these stages as the moisture content changes.

1. *Liquid Limit*. During the drying process, the initially liquid state reaches a consistency at which the soil ceases to behave as a liquid and begins to exhibit the behavior of a plastic (rapid deformation but no fracture under constant load, no rebound if unloaded). The water content at this stage is called the liquid limit (LL).
2. *Plastic Limit*. As the drying process continues, the plastic state reaches a consistency at which the soil ceases to behave as a plastic and begins to break apart and crumble when rolled by hands into cylinders $\frac{1}{8}$ in. in diameter. The water content at this stage is called the plastic limit (PL).
3. *Shrinkage Limit*. Finally, the water content at which the soil ceases to shrink is denoted as the shrinkage limit (SL).

Index Properties

Atterberg's limits alone are of limited practical value, but they are very helpful in the process of identification and classification of soils. Three index definitions are based on these limits:

1. *Plasticity Index* (PI) is the difference between the liquid limit (LL) and the plastic limit (PL).
2. *Shrinkage Index* (SI) is the difference between the plastic limit (PL) and the shrinkage limit (SL).
3. *Liquidity Index* (LI) is given as

$$(\text{LI}) = \frac{w - (\text{PL})}{(\text{LL}) - (\text{PL})} = \frac{w - (\text{PL})}{(\text{PI})} \qquad (1\text{-}11)$$

and is negative if (PL) $>$ w. The value of (LI) increases from 0.0 to 1.0 as the water content increases from the plastic limit to the liquid limit (Problems 1.10, 1.11, and 1.12).

The diagrammatic representation of four consistencies, three Atterberg's limits, and two related indexes are shown in Table 1-1 (Problems 1.12 through 1.15).

Atterberg's Systems, Limits, and Indexes   **TABLE 1-1**

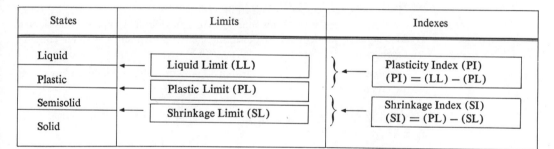

States	Limits	Indexes
Liquid Plastic Semisolid Solid	Liquid Limit (LL) Plastic Limit (PL) Shrinkage Limit (SL)	Plasticity Index (PI) (PI) = (LL) − (PL) Shrinkage Index (SI) (SI) = (PL) − (SL)

SOLVED PROBLEMS

Definition of Terms

1.1 State the engineering definition of soil and compare it with the geological definition.

For engineering purposes, soil is considered a natural aggregate of mineral grains, *Answer*
with or without organic constituents, that can be separated by gentle mechanical
means, such as agitation in water. It is regarded as any loose sedimentary or residual
deposit, such as gravel, sand, clay, or a mixture of these materials. It is not to be
confused with the geological definition of soil, which is the weathered organic
material on the surface, or topsoil. Topsoil is generally removed before any engineering project is carried out.

1.2 Define the weathering process, plus its special forms, and give typical examples.

Answer The processes of weathering are subdivided into the following special forms:

1. *Disintegration*, which refers to the weathering of rock by physical agents, such as
   (a) Periodical temperature changes, which are especially pronounced in deserts, where days are hot and nights are cold. The rock in such cases fails by fatigue because of the continuous reversal of compression and tensile stresses produced in it by the temperature changes.
   (b) Freezing and thawing, as a result of the expansion of water inside the rock pores when it freezes and the rock fails in tension.
   (c) Physical effects of plants and animal life on rock.
2. *Decomposition*, which refers to the weathering of rock by chemical agents, such as
   (a) Oxidation, which means that oxygen ions are added to the minerals composing the rock. Rocks containing iron, for example, are greatly affected by oxidation.
   (b) Reduction, which is the opposite of oxidation, meaning that oxygen ions are removed from the minerals in rock.
   (c) Hydration is usually the chemical addition of water to minerals and should be distinguished from the free water that causes disintegration. An important hydration effect is the decomposition of the feldspar mineral in granite to form the clay mineral kaolinite.
3. *Solution* or *carbonation*, which is the solution of the rock material by water containing a considerable amount of carbon dioxide. This process is considered detrimental, particularly in the limestone areas. All surface water contains carbon dioxide.

These weathering processes may occur individually or collectively, depending on the climatic conditions.

1.3 Describe the composition of organic matter in soil and its significance in engineering.

Answer The organic matter in soil is derived from either plant or animal remains that are added to the soil when the organisms die and that subsequently undergo decomposition due to chemical and bacterial action. Under normal conditions, they tend to be concentrated in the top 2 to 12 in. of the soil profile. From the engineering point of view, organic matter has undesirable characteristics resulting from the open spongy structure and the mechanical weakness of the constituents. It undergoes considerable volume changes when subjected to load or moisture changes.

Organic soils have a high natural moisture content (100 to 500 percent) and their mechanical stability is low. The acidic nature of the constituents tends to give an acid reaction to the water in soil, which in turn may have a corrosive effect on materials buried in soil. Shallow deposits of soils containing appreciable amounts of organic matter are generally removed from a site prior to construction.

1.4 Define a rock mineral and list the most important rock-forming minerals.

Answer Geologically, the word mineral means a naturally occurring inorganic substance with a characteristic internal structure and with a chemical composition and physical properties that are either uniform or variable within definite limits. Most minerals are compounds of two or more elements, but a few, such as sulfur and gold, are single elements.

The rock-forming minerals are normally classified on the basis of two categories.

1. The nature of the atoms. According to this classification, minerals are classified as carbonates, phosphates, oxides, and silicates. This classification is of limited value to civil engineers, for the most important and abundant minerals are silicates.

2. The arrangement of atoms. This classification is more important, for there is a relationship between the atomic arrangement in a mineral and its physical, optical, chemical, and engineering properties. The earth's crust is made up primarily of nine minerals: feldspars (orthoclase, microcline, and plagioclase), olivine, pyroxenes, amphiboles, micas (biotite and muscovite), and quartz. They differ from each other in the silicate structure. For example, the olivine mineral is composed of independent tetrahedrons with no shared oxygens per silicon tetrahedron, which makes it an unstable mineral under the action of weathering. Quartz mineral, on the other hand, has a framework of silicate structure with four oxygens shared per tetrahedron, which makes it one of the most weather-resistant minerals.

1.5   List the most important rock minerals forming the solids in soils and give brief descriptions of them.

The most common rock minerals in soils are   *Answer*

1. *Quartz* (SiO_2). Clear transparent mineral formed by the slow cooling of liquid magma. Because of its high resistance to chemical weathering, it breaks into smaller particles by mechanical weathering without any change in its composition. It constitutes the principal mineral in sands, silts, and rock flour and is No. 7 in the rock-hardness scale.

2. *Feldspars.* Generally light colored and characterized by two good cleavages. Feldspars are No. 6 on the rock-hardness scale. They are divided into three types (Problem 1.2), depending on composition, cleavage, and crystal structure. The two important types are
   (a) *Orthoclase* (from the Greek *ortho*, straight + *klasis*, fracture), in which cleavages form angles of 90 degrees. Its chemical formula is $K_2O \cdot Al_2O_3 \cdot 6SiO_2$. Sands deposited by rivers in many parts of the world contain considerable amounts of orthoclase feldspar.
   (b) *Plagioclase* (from the Greek *plagios*, oblique + *klasis*, fracture), in which cleavages make angles of about 86 degrees. Its chemical formula is $Na_2O \cdot Al_2O_3 \cdot 6SiO_2$. Sands of ferromagnesian minerals contain plagioclase feldspars.

3. *Micas.* There are two varieties.
   (a) Muscovite, hydrous potassium aluminum silicate, $KAl_3 \cdot Si_3O_{10} \cdot (OH)_2$, which is white or colorless.
   (b) Biotite, $K(Mg \cdot Fe)_3Al \cdot Si_3O_{10} \cdot (OH)_2$, which is black. Sand-sized particles of mica are often found in sands, and colloidal sizes occur in clays.

1.6   List the most important clay minerals forming the solids in soils and give brief descriptions of them.

Answer The most common clay minerals in soils are

1. *Kaolinite*, hydrous aluminum silicate, $Al_2Si_2O_5 \cdot (OH)_4$. A light-colored mineral in minute particles, derived from the chemical weathering of feldspars as follows:

$$2KAlSi_3O_8 + 2H_2O + CO_2 \rightarrow H_4Al_2Si_2O_9 + K_2CO_3 + 4SiO_2$$

It is the main constituent of many clay soils of humid-temperate and humid-tropical regions. It occurs in clay as platelets from 1000 to 20,000 Å wide × 100 to 1000 Å thick. It feels greasy and is plastic when wet. It consists of alternate layers of silica and gibbsite [1Å (angstrom) = 1×10^{-4} mm].

2. *Montmorillonite*, hydrous aluminum silicate, $Al_2O_35SiO_2 \cdot 5\text{-}7H_2O$. Most prominent in clay soils of arid regions—for example, deserts, prairies with low rainfall. It is more colloidal than kaolinites and swells considerably on wetting. Figure P-1.6 illustrates the thickness of soil moisture on clay particles of typical dimensions.

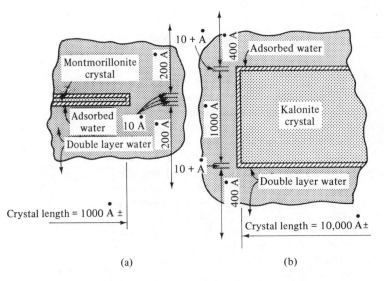

3. Illite is a hydrous aluminum silicate similar to montmorillonite except that the silica layers are bonded with potassium ions instead of water. It is also known by the term hydrous micas. Illite is the main constituent of many clay shales. It is more active than kaolinite with water but much less active than montmorillonite. Lateral dimensions of illite particles are the same as montmorillonite, but the thickness is greater, about 50 to 500 Å.

Basic Properties

1.7   Prove the following relationships:

(I) $\gamma_d = \dfrac{\gamma_m}{1+w}$ (II) $n = 1 - \dfrac{\gamma_m}{G_s\gamma_0(1+w)}$

The proofs presented below are based on the relationships of Section 1.3 referred   *Solution*
to in Fig. 1-2a.

I.   Dry unit weight (density) (1-7):

$$\frac{\gamma_m}{1+w} = \left(\frac{W_s+W_w}{V_m}\right)\left[\frac{1}{1+(W_w/W_s)}\right] = \left(\frac{W_s+W_w}{V_m}\right)\left(\frac{W_s}{W_s+W_w}\right)$$

$$= \frac{W_s}{V_m} = \gamma_d \tag{a}$$

where γ_d is the oven-dry unit weight of soil (1-7).

II.   Porosity (1-3):

$$n = \frac{V_v}{V_m} = \frac{V_m - V_s}{V_m} \tag{b}$$

where by (1-9)

$$V_s = \frac{W_s}{G_s\gamma_0} \tag{c}$$

Then (b) in terms of (c) is

$$n = \frac{V_m - (W_s/G_s\gamma_0)}{V_m} = 1 - \frac{W_s}{V_mG_s\gamma_0} \tag{d}$$

which by (1-7) reduces to

$$n = 1 - \frac{\gamma_d}{G_s\gamma_0} \tag{e}$$

However, it was shown in equation (a) of this problem that

$$\gamma_d = \frac{\gamma_m}{1+w} \tag{f}$$

with which (e) yields the required relationship

$$n = 1 - \frac{\gamma_m}{G_s\gamma_0(1+w)} \tag{g}$$

1.8   Prove the following relationships:

(I) $S = \dfrac{wG_s(1-n)}{n}$ (II) $\gamma_d = \dfrac{G_s}{1+e}\gamma_w$

Solution As in Problem 1.7, the subsequent proofs are based on the relationships of Section 1.3 referred to in Fig. 1-2a.

I. Degree of saturation (1-5):

$$S = \frac{V_m}{V_v} = \frac{V_w/V_s}{e} = \frac{wG_s}{e} \tag{a}$$

and in terms of (1-4)

$$S = \frac{wG_s(1-n)}{n} \tag{b}$$

II. Dry unit weight (1-7):

$$\gamma_d = \frac{W_s}{V_m} = \frac{G_s V_s \gamma_w}{V_m} \tag{c}$$

where

$$W_s = G_s V_s \gamma_w$$

Then in terms of (1-4)

$$e = \frac{V_v}{V_s}$$

and the dry unit weight (c) becomes

$$\gamma_d = \frac{G_s}{1+e} \gamma_w \tag{d}$$

1.9 A sample of saturated soil has a unit weight $\gamma_m = 130 \text{ lb/ft}^3$ and a water content $w = 25$ percent. Calculate the porosity n and the void ratio e of this soil. Assume that $\gamma_w = \gamma_0 = 62.4 \text{ lb/ft}^3$.

Solution The total weight of 1 ft³ of this soil, represented by the block diagram of Fig. P-1.9 in terms of (1-2), is

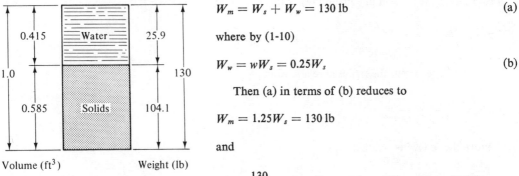

$$W_m = W_s + W_w = 130 \text{ lb} \tag{a}$$

where by (1-10)

$$W_w = wW_s = 0.25W_s \tag{b}$$

Then (a) in terms of (b) reduces to

$$W_m = 1.25W_s = 130 \text{ lb}$$

and

$$W_s = \frac{130}{1.25} = 104.1 \text{ lb} \qquad W_w = W_m - W_s = 25.9 \text{ lb}$$

Volume (ft³)   Weight (lb)

Figure P-1.9

The volume of water (1-7) is

$$V_w = \frac{W_w}{\gamma_0} = \frac{25.9}{62.4} = 0.415 \text{ ft}^3$$

and the volume of solids (1-1) is

$$V_s = V_m - V_w = 1.000 - 0.415 = 0.585 \text{ ft}^3$$

since $V_g \cong 0$.

Finally, the three typical volume ratios—the porosity n (1-3), the void ratio e (1-4), and the degree of saturation S (1-5)—are, respectively,

$$n = \frac{V_v}{V_m} = \frac{V_w}{V_m} = \frac{0.415}{1.000} = 0.415$$

$$e = \frac{V_v}{V_s} = \frac{V_w}{V_s} = \frac{0.415}{0.585} = 0.715$$

$$S = \frac{V_w}{V_v} = \frac{V_w}{V_w} = \frac{0.415}{0.415} = 1.000 \text{ or } 100 \%$$

where the equality $V_w = V_v$ is based on the condition of full saturation (all voids are filled with water).

1.10   For the sample of soil given by the total weight $W_m = 68$ lb, the total volume $V_m = 0.660 \text{ ft}^3$, the dry weight $W_d = 60$ lb, and the specific gravity of solids $G_s = 2.5$, compute

(a)   the unit weight of the sample γ_m.
(b)   the dry unit weight (dry density) γ_d.
(c)   the water content w.
(d)   the void ratio e.
(e)   the porosity n.
(f)   the degree of saturation S.

The results are calculated by the equations identified in the parentheses below and *Solution* by the block diagram of Fig. P-1.10.

(a)   Unit weight (1-6) is

$$\gamma_m = \frac{W_m}{V_m} = \frac{68}{0.660} = 103 \text{ lb/ft}^3$$

(b)   Dry density (1-7) is

$$\gamma_d = \frac{W_s}{V_m} = \frac{60}{0.660} = 91 \text{ lb/ft}^3$$

(c)   Water content (1-10) is

$$w = \frac{W_w}{W_s} = \frac{8}{68} = 0.133$$

where

$$W_w = W_m - W_s = 68 - 60 = 8 \text{ lb}$$

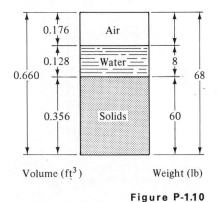

Volume (ft³)   Weight (lb)

Figure P-1.10

(d) Void ratio (1-4) is

$$e = \frac{V_v}{V_s} = \frac{0.304}{0.356} = 0.851$$

where by (1-9)

$$V_s = \frac{W_s}{G_s \gamma_0} = \frac{60}{(2.70)(62.4)} = 0.356 \text{ ft}^3$$

and by (1-1)

$$V_v = V_m - V_s = 0.660 - 0.356 = 0.304 \text{ ft}^3$$

(e) Porosity (1-3) is

$$n = \frac{V_v}{V_m} = \frac{0.304}{0.660} = 0.461$$

(f) Degree of saturation (1-5) is

$$S = \frac{V_w}{V_v} = \frac{0.128}{0.304} = 0.422$$

1.11 In a test to determine the *in situ* density of a soil, a sample of 10.5 lb of soil was excavated from a hole at the surface of the ground. In order to measure the volume of the excavated soil, the same hole was then filled with 8.05 lb of loose dry sand of unit weight 96.6 lb/ft.^{3} If 25 g of the excavated soil weighed 21 g after drying in the laboratory oven and the specific gravity of its solids was found to be $\gamma_s = 2.69$, compute

(a) the volume of the excavated soil V_m,
(b) the in situ unit weight γ_m,
(c) the in situ (natural) water content w,
(d) the dry density γ_d,
(e) the degree of saturation S.

Solution The following calculations refer to the block diagram of Fig. P-1.11.

(a) Volume of the excavated hole is

$$V_m = \frac{8.05}{96.6} = 0.0832 \text{ ft}^3$$

(b) In situ unit weight (1-7) is

$$\gamma_m = \frac{10.5}{0.0832} = 126 \text{ lb/ft}^3 = 2.02 \text{ g/cm}^3$$

(c) Natural moisture content (1-10) is

$$w = \frac{4}{21} = 0.19 \text{ or } 19\%$$

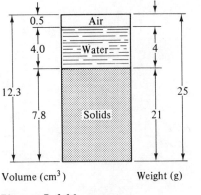

Volume (cm^3)   Weight (g)

Figure P-1.11

(d) Dry density (1-7) is

$$\gamma_d = \frac{\gamma_m}{1+w} = \frac{126}{1+0.19} = 106 \text{ lb/ft}^3$$

(e) $V_m = \dfrac{W_m}{\gamma_m} = \dfrac{25}{2.02} = 12.3 \text{ cm}^3 \qquad V_s = \dfrac{W_s}{G_s\gamma_0} = \dfrac{21}{(2.69)(1.0)} = 7.8 \text{ cm}^3$

$V_v = V_m - V_s = 12.3 - 7.8 = 4.5 \text{ cm}^3 \qquad V_w = 4 \text{ cm}^3$

and degree of saturation (1-5) is

$$S = \frac{4}{4.5} = 0.89 \text{ or } 89 \%$$

Soil Consistency

1.12 Describe the testing procedure and the apparatus for determining the liquid limit (LL) of clay.

The apparatus for the determination of this limit is shown in Fig. P-1.12.   *Answer*

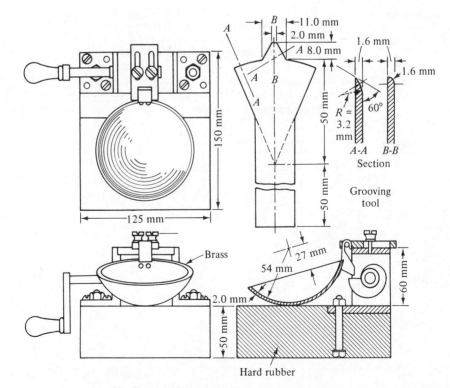

(A. Casagrande, "Research of the Atterberg's Limits of Soils," *Public Roads*, Vol. 13, No. 8, October 1932, pp. 121-136.)   **Figure P-1.12**

For its application, the following procedure is used:

1. A 100 g of moist soil is mixed with distilled water until a uniform paste is obtained.

2. Then a part of this paste sample is placed in the cup of the apparatus to form a smooth surface of a maximum depth of $\frac{1}{2}$ in.

3. Using the grooving tool, a cut is made through the paste along the axis of symmetry normal to the axis of the crank.

4. By turning the crank at about 2 rps (revolutions per second) and counting the blows (rotations), the procedure will eventually close the groove for a distance of $\frac{1}{2}$ in.

The repetition of this procedure at several water contents allows the construction of a plot relating the water content to the number of blows. The liquid limit (LL) is then the water content at 25 blows.

1.13 If the liquid limit test described in Problem 1.12 yielded the following results,

Test trial number	1	2	3	4
Number of blows	12	19	27	40
Moisture content	54.5	51.8	49.5	47

what is the liquid limit of the tested sample ?

Answer The liquid limit test data are plotted on a semilog scale as shown in Fig. P-1.13. The liquid limit of the given soil is then scaled off from the blow graph at 25 blows and in this case is 50 percent.

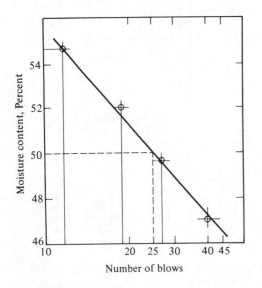

Figure P-1.13

1.14   Discuss the liquid limit of various types of soils.

The liquid limit (LL) of noncohesive soils, such as silty sands and sands, is usually   *Answer*
less than 25 percent. On the other hand, LL of clayey and silty soils may reach up
to 100 percent. An average value of LL of cohesive soils is about 60 percent. When
LL is high, it indicates that the soil has a large percent of clay.

1.15   The following results were obtained by testing a sample of clay: $w = 25.6\%$,
$LL = 43.5\%$, $PL = 19.1\%$. Compute the plasticity index (PI) and the liquidity
index (LI).

Plasticity Index   *Solution*

$$PI = LL - PL = 43.5 - 19.1 = 24.4\%$$

Liquidity Index

$$LI = \frac{w - PL}{PI} = \frac{25.6 - 19.1}{24.4} = 0.267$$

It must be noted that the significance of the LI lies in the fact that the LL and
PL tests can only be performed on a remolded soil. At the same water content, a
soil in the natural or undisturbed state may exhibit a different consistency. Therefore
a useful index that reflects the properties of natural soils is the liquidity index (LI).

Soil Structure

1.16
(a)   Describe the ideal cubic and ideal rhombic packing of granular particles.
(b)   What is the void ratio for ideal cubic and ideal rhombic packings ?

(a)   Figure P-1.16a shows a plan and vertical view of simple cubic packing; Fig.   *Answer*

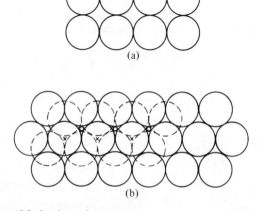

(a)

(b)

(H. Deresiewicz, "Mechanics of Granular   **Figure P-1.16**
Matter," *Advances in Applied Mechanics*, Vol.
5, New York, Academic Press, 1958, p. 233.)

P-1.16b shows a plan view of dense or ideal rhombic packing. Solid circles, first layer; dashed circles, second layer; 0 location of sphere centers in third layer.

(b) Maximum and minimum densities of granular soils (depending on the degree of packing of the grains) are given in Table P-1.16.

TABLE P-1.16

Description	VOID RATIO		POROSITY (%)		DRY UNIT WEIGHT (PCF)	
	e_{max}	e_{min}	n_{max}	n_{min}	$\gamma_{d\ min}$	$\gamma_{d\ max}$
Uniform spheres	0.92	0.35	47.6	26.0	–	–
Standard Ottawa sand	0.80	0.50	44	33	92	110
Clean uniform sand	1.0	0.40	50	29	83	118
Uniform inorganic silt	1.1	0.40	52	29	80	118
Silty sand	0.90	0.30	47	23	87	127
Fine-to-coarse sand	0.95	0.20	49	17	85	138
Micaceous sand	1.2	0.40	55	29	76	120
Silty sand and gravel	0.85	0.14	46	12	89	146

K. B. Hough, Basic Soil Engineering, copyright 1957, The Ronald Press, New York, pp. 30-31.

1.17 What is the density index (measure) of the structure of natural granular soil ? Give some examples.

Answer The most common expression used for the measuring of this characteristic is the relative density, D_r.

$$D_r = \left(\frac{e_{max} - e}{e_{max} - e_{min}}\right) 100\% = \left(\frac{\gamma_{d\ max}}{\gamma_d}\right)\left(\frac{\gamma_d - \gamma_{d\ min}}{\gamma_{d\ max} - \gamma_{d\ min}}\right) 100\%$$

where e_{min} = void ratio of soil in densest condition,

e_{max} = void ratio of soil in loosest condition,

e = in-place void ratio,

$\gamma_{d\ max}$ = dry unit weight of soil in densest condition,

$\gamma_{d\ min}$ = dry unit weight of soil in loosest condition,

γ_d = in-place dry unit weight

Some typical examples of D_r are given below.

Structure	Very Loose	Loose	Medium	Dense	Very Dense
$D_r \%$	0–15	15–35	35–65	65–85	85–100

SUPPLEMENTARY PROBLEMS

Definition of Terms

1.18 Arrange common rock minerals in order of their resistance to weathering.

Answer

TABLE P-1.18

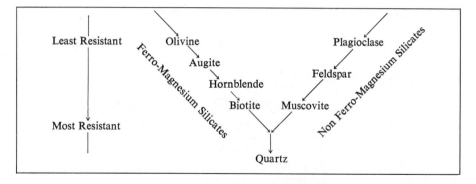

NOTE: This arrangement should not infer that olivine weathers to augite or augite to hornblende, but rather that in a rock containing both olivine and augite that olivine will weather more rapidly than augite. It should be clear from this table why quartz and muscovite are common constituents of weathered rock residues.

1.19 Explain very briefly the cause of the difference in volume changes that occur among the three common types of clay minerals in the presence of water.

Refer to Fig. P-1.6. Different types of clay minerals differ in their chemical   *Answer*
composition and their electrical surface charge. More active minerals (e.g., montmorillonite) attract thicker moisture films and tend to swell and shrink more with changes in moisture content than other minerals that are less active. In the case of montmorillonite, there is also very little bonding force between the successive sheets that make up the individual clay particles. Water, when available, enters between these sheets, causing even more swelling.

1.20   What clay mineral would probably be most common in the following soils ?
(a)   Desert soil.
(b)   Laterite.
(c)   Old glacial till (e.g., Illinois).
(d)   Loess.
(e)   A mechanically weathered soil residually developed on a biotite gneiss (a weathered rock).

Answer   (a)   Montmorillonite.   (b)   Kaolinite.   (c)   Illite.   (d)   Illite.   (e)   Illite.

1.21   In each part of this problem, data are presented or comparisons drawn between the three clay minerals: illite (I), montmorillonite (M), and kaolinite (K). Place the proper symbol beside the properties given below.

(a)   Plasticity Index (PI) is 29
Plasticity Index (PI) is 350
Plasticity Index (PI) is 14
(b)   Highest dry strength
Lowest dry strength
Intermediate dry strength

(c)   Largest wet-dry volume change
Intermediate value
Lowest value
(d)   Coarsest grain size
Finest grain size
Intermediate grain size

Answer   (a)   I, M, K   (b)   M, K, I   (c)   M, I, K   (d)   K, M, I

Basic Properties

1.22   A sample of wet soil weighs 37.6 grams. The sample is coated with wax (density 0.90 g/cm^3) and then weighs 40.9 g. The wax-coated sample is then weighed immersed in water and found to be 16.5 g. If the water content of the sample w is 17.4 percent and the specific gravity G_s is 2.70, what is the
(a)   wet density,
(b)   dry density,
(c)   porosity,
(d)   percentage saturation ?

Answer   (a) $\gamma_m = 113.20 \, \text{lb/ft}^3$
(b) $\gamma_d = 96.40 \, \text{lb/ft}^3$
(c) $n = 0.428$
(d) $S = 62.8 \%$

1.23   In the case of a saturated soil sample, there are certain algebraic relations between $\gamma_{\text{sat}}, \gamma_d. \gamma_0, G_s, n, e, w$, and the analytical expressions for $G_s, \gamma_d, \gamma_{\text{sat}}, w, n$, or e can be formulated in terms of any three of the remaining quantities. State these relationships. Note that $\gamma_{\text{sat}} = \gamma_m$ in this special case.

Answer   Table P-1.23

Functional Relationships   TABLE P-1.23

Given Quantities	Specific Gravity G_s	Dry Unit Weight γ_d	Saturated Unit Weight γ_{sat}	Saturated Moisture Content $w\%$	Porosity n	Void Ratio e
γ_0, G_s, γ_d	–	–	$\left(1-\frac{1}{G_s}\right)\gamma_d+\gamma_0$	$\left(\frac{1}{\gamma_d}-\frac{1}{G\gamma_0}\right)\gamma_0$	$1-\frac{\gamma_d}{G\gamma_0}$	$\frac{G\gamma_0}{\gamma_d}-1$
$\gamma_0, G_s, \gamma_{sat}$	–	$\frac{\gamma_{sat}-\gamma_0}{G_s-1}G_s$	–	$\frac{G_s\gamma_0-\gamma_{sat}}{(\gamma_{sat}-\gamma_w)G_s}$	$\frac{G_s\gamma_0-\gamma_{sat}}{(G_s-1)\gamma_0}$	$\frac{G_s\gamma_0-\gamma_{sat}}{\gamma_{sat}-\gamma_0}$
γ_0, G_s, W	–	$\frac{G_s}{1+wG_s}\gamma_0$	$\frac{1+w}{1+wG_s}G_s\gamma_0$	–	$\frac{wG_s}{1+wG_s}$	wG_s
γ_0, G_s, n	–	$G(1-n)\gamma_0$	$[G_s-n(G_s-1)]\gamma_0$	$\frac{n}{G_s(1-n)}$	–	$\frac{n}{1-n}$
γ_0, G_s, e	–	$\frac{G_s}{1+e}\gamma_0$	$\frac{G_s+e}{1+e}\gamma_0$	$\frac{e}{G_s}$	$\frac{e}{1+e}$	–
$\gamma_0, \gamma_d, \gamma_{sat}$	$\frac{\gamma_d}{\gamma_0+\gamma_d-\gamma_{sat}}$	–	–	$\frac{\gamma_{sat}}{\gamma_d}-1$	$\frac{\gamma_{sat}-\gamma_d}{\gamma_0}$	$\frac{\gamma_{sat}-\gamma_d}{\gamma_0+\gamma_d-\gamma_{sat}}$
γ_0, γ_d, w	$\frac{\gamma_d}{\gamma_0-w\gamma_d}$	–	$(1+w)\gamma_d$	–	$w\frac{\gamma_d}{\gamma_w}$	$\frac{w\gamma_d}{\gamma_0-w\gamma_d}$
γ_0, γ_d, n	$\frac{\gamma_d}{(1-n)\gamma_0}$	–	$\gamma_d+n\gamma_0$	$\frac{n\gamma_0}{\gamma_d}$	–	$\frac{n}{1-n}$
γ_0, γ_d, e	$(1+e)\frac{\gamma_d}{\gamma_0}$	–	$\frac{e\gamma_0}{1+e}+\gamma_d$	$\frac{e}{1+e}\left(\frac{\gamma_0}{\gamma_d}\right)$	$\frac{e}{1+e}$	–
$\gamma_0, \gamma_{sat}, w$	$\frac{\gamma_{sat}}{\gamma_0-w(\gamma_{sat}-\gamma_0)}$	$\frac{\gamma_{sat}}{1+w}$	–	–	$\frac{w\gamma_{sat}}{(1+w)\gamma_0}$	$\frac{w\gamma_{sat}}{\gamma_0-w(\gamma_{sat}-\gamma_0)}$
$\gamma_0, \gamma_{sat}, n$	$\frac{\gamma_{sat}}{\gamma_0(1+w)(1-n)}$	$\gamma_{sat}-n\gamma_0$	–	$\frac{n\gamma_0}{\gamma_{sat}-n\gamma_0}$	–	$\frac{n}{1-n}$
$\gamma_0, \gamma_{sat}, e$	$(1+e)\frac{\gamma_{sat}}{\gamma_0}-e$	$\gamma_{sat}-\frac{e}{1+e}\gamma_0$	–	$\frac{e\gamma_0}{\gamma_{sat}+e(\gamma_{sat}-\gamma_0)}$	$\frac{e}{1+e}$	–
γ_0, w, n	$\frac{n}{(1-n)w}$	$\frac{n}{w}\gamma_0$	$n\left(\frac{1+w}{w}\right)\gamma_0$	–	–	$\frac{n}{1-n}$
γ_0, w, e	$\frac{e}{w}$	$\frac{e}{(1-e)w}\gamma_0$	$\frac{e}{w}\left(\frac{1+w}{1+e}\right)\gamma_0$	–	$\frac{e}{1+e}$	–

From Soil Mechanics *by A. R. Jumikis, ©1962 by Litton Educational Publishing, Inc. Reprinted by permission of Van Nostrand Reinhold Company.*

1.24 A soil with a liquidity index (LI) $= -0.16$ has a liquid limit of 37.5 percent and a plasticity index (PI) of 1.32 percent. What is the natural water content of this soil?

$w = 22.19\%$ *Answer*

1.25 A tropical laterite soil known to be an extrasensitive clay was found to have an unconfined compressive strength of $0.27\ t/ft^2$. What would be the minimum probable value of the undisturbed strength of this clay?

Answer $2.16\ t/ft^2$

1.26 In a compaction test, the weight of wet soil in a standard compaction mould (volume $\frac{1}{30}$ ft³) was 4.23 lb. After drying a small quantity of the wet soil in an oven, the moisture content was found to be 21.0 percent. The specific gravity of the solid particles was $G_s = 2.71$. Compute
(a) the dry density.
(b) the void ratio.
(c) percent air voids.
(d) the moisture content of this soil if completely saturated.

Answer (a) $\gamma_{dry} = 104.9\ lb/ft^3$ (b) $e = 0.6125$ (c) 2.73 % (d) 22.6 %

Soil Consistency

1.27
(a) Define the liquid limit (LL) and the plasticity index (PI) of soil.
(b) Explain their significance as indicators of the engineering characteristics of the soil.

Answer (a) Refer to Section 1.5.
(b) Significance of LL and PI: high values of LL and PI indicate that the soil
(1) has a high percent of clay and colloidal sizes of the more-active-type minerals, (2) has high resiliency, thus making it difficult to compact as a highway subgrade, (3) possesses a poor foundation or traffic load-bearing capacity, and (4) is more susceptible to volume change with moisture fluctuations, thus making it an undesirable foundation for engineering structures, especially pavements.

1.28 A sample of saturated soil had a volume of 100 cm³ and weighed 206 grams. When the sample was completely dried out, the volume of the sample was 89 cm³ and its weight 165 g. Compute
(a) the initial moisture content.
(b) shrinkage limit (SL).
(c) the specific gravity of the soil G_m.

Answer (a) $w = 2.48\ \%$ *24.8 %*
(b) $SL = w - \dfrac{(\Delta V)\gamma_0}{W_s} \times 100$
where $(\Delta V) = V_0 - V_f$
V_0 = volume of wet soil
V_f = volume of oven-dried soil
$SL = 18.1\ \%$

(c) $G_m = \dfrac{165}{89} = 1.86$

classification

2

Purpose

The principal objectives in the engineering classification of soils are the establishment of a *common language* for technical communication and the *arrangement of soils into groups* based on their engineering applications. Although several classification systems have been devised, none of these systems can adequately describe any soil for all possible engineering applications and, consequently, none is completely satisfactory.

Categories

The soil classification systems fall into three major categories: (a) classification by geological and pedological criteria, (b) classification by morphological properties, and (c) classification by use.

2.2 GEOLOGICAL AND PEDOLOGICAL CLASSIFICATIONS

Classification by Origin

The classification by origin is mostly genetic but also descriptive. Two major divisions of this classification are shown on the following page:

25

1. *Residual soils*, such as zonal, azonal, and intrazonal mineral conglomerates, are soils located at the place of their formation (nontransported soils) (Problem 2.1).
2. *Transported soils*, such as fluvial, alluvial, aeolian, glacial, marine, volcanic deposits, and mass-wasting dislocations (slope wash, slide rocks, mudflows, etc.), are soils that have been transported from their place of origin by water, ice, wind and/or gravity (Problem 2.2).

Since the structure and behavior of soils are governed to some extent by their origin and formation, this classification is important in engineering.

Pedological Classification

The systematic investigation of the behavior of topsoils is called the *pedology*. The pedological classification is related to the *soil profile* (vertical cross section), which is influenced by five principal genetic factors: (a) *climate*, (b) *vegetation*, (c) *parent material*, (d) *topography*, and (e) *age*.

The degree of participation of these factors produces the particular soil, which is then classified by *order*, *suborder*, *great soil group*, *soil series*, and *soil type*. The introduction of the pedological classification in engineering soil mechanics is of recent date, and the new trends indicate some promising possibilities (Problems 2.3, 2.4, and 2.5).

2.3 MORPHOLOGICAL CLASSIFICATIONS

Classification by Appearance

The macroscopic classification, based on visual and manual inspection, has already been introduced in Section 1.2 and illustrated by Fig. 1-1. A set of guidelines for the field identification is given in Table 2-1 in this section.

Grain-Size Classification

Since the grain size is one of the most apparent physical characteristics of soil, it formed the basis of the earlier classification systems. Six typical classification charts based on the grain-size distribution are shown in Table 2-3. For their application, the sieve analysis must be performed first, and the sample must be segregated into groups. This process is accomplished by means of standard sieves, the sizes and designations of which are given in Table 2-2. Then the weight of each segregation is expressed as a percentage of the total weight of the sample (Problems 2.6 and 2.7) and the soil is designated by the class of the respective chart.

Soil Texture	Visual Detection of Particle Size and General Appearance of the Soil	SQUEEZED IN HAND AND PRESSURE RELEASED		Soil ribboned between thumb and finger when moist.
		When Air Dry	When Moist	
Sand	Soil has a granular appearance in which the individual grain sizes can be detected. It is free-flowing when in a dry condition.	Will not form a cast and will fall apart when pressure is released.	Forms a cast that will crumble when lightly touched.	Cannot be ribboned.
Sandy loam	Essentially a granular soil with sufficient silt and clay to make it somewhat coherent. Sand characteristics predominate.	Forms a cast that readily falls apart when lightly touched.	Forms a cast that will bear careful handling breaking.	Cannot be ribboned.
Loam	A uniform mixture of sand, silt and clay. Grading of sand fraction quite uniform from coarse to fine. It is mellow, has somewhat gritty feel, yet is fairly smooth and slightly plastic.	Forms a cast that will bear careful handling without breaking.	Forms a cast that can be handled freely without breaking.	Cannot be ribboned.
Silt loam	Contains a moderate amount of the finer grades of sand and only a small amount of clay, over half of the particles are silt. When dry, it may appear quite cloddy, which can be readily broken and pulverized to a powder.	Forms a cast that can be freely handled. Pulverized it has a soft, flour-like feel.	Forms a cast that can be freely handled. When wet, soil runs together and puddles.	It will not ribbon, but it has a broken appearance, feels smooth, and may be slightly plastic.
Silt	Contains over 80 % of silt particles with very little fine sand and clay. When dry, it may be cloddy, readily pulverizes to powder with a soft flourlike feel.	Forms a cast that can be handled without breaking.	Forms a cast that can be handled freely. When wet, it readily puddles.	It has a tendency to ribbon with a broken appearance, feels smooth.
Clay loam	Fine textured soil breaks into hard lumps when dry. Contains more clay than silt loam. Resembles clay in a dry condition; identification is made on physical behavior of moist soil.	Forms a cast that can be freely handled without breaking.	Forms a cast that can be handled freely without breaking. It can be worked into a dense mass.	Forms a thin ribbon that readily breaks, barely sustaining its own weight.
Clay	Fine-textured soil breaks into very hard lumps when dry. Difficult to pulverize into a soft flourlike powder when dry. Identification based on cohesive properties of the moist soil.	Forms a cast that can be freely handled without breaking.	Forms a cast that can be handled freely without breaking.	Forms long, thin, flexible ribbons. Can be worked into a dense, compact mass. Considerable plasticity.
Organic soils	Identification based on the high organic content. Muck consists of thoroughly decomposed organic material with considerable amount of mineral soil finely divided with some fibrous remains. When considerable fibrous material is present, it may be classified as peat. The plant remains, or sometimes the woody structure can easily be recognized. Soil color ranges from brown to black. They occur in lowlands, in swamps or swales. They have high shrinkage upon drying.			

TABLE 2-2   *Standard Sieve Sizes*

U.S. STANDARD			BRITISH STANDARD			METRIC STANDARD		
No.	D (mm)	D (in.)	No.	D (mm)	D (in.)	No.	D (mm)	D (in.)
4	4.76	0.1874	5	3.36	0.1323	5000	5.00	0.1969
6	3.36	0.1323	8	2.06	0.0811	3000	3.00	0.1181
10	2.00	0.0787	12	1.41	0.0555	2000	2.00	0.0787
20	0.84	0.0331	18	0.85	0.0335	1500	1.54	0.0606
40	0.42	0.0165	25	0.60	0.0236	1000	1.00	0.0394
60	0.25	0.0098	36	0.42	0.0165	500	1.00	0.0197
100	0.149	0.0059	60	0.25	0.0098	300	0.50	0.0118
200	0.074	0.0029	100	0.15	0.0059	150	0.15	0.0059
			200	0.076	0.0030	75	0.075	0.0030

TABLE 2-3   *Grain-Size Classification*

System*	Grain Diameter in millimeters

(a)

0.0002	0.0006	0.002	0.006	0.02	0.06	0.2	0.6	2.0 mm
Colloids	Medium	Coarse	Fine	Medium	Coarse	Fine	Medium	Coarse	Gravel
Clay			Silt			Sand		

(b)

0.001	0.005	0.074	0.25	2.0 mm
Colloids	Silt	Sand	Fine / Coarse (Sand)	Gravel

(c)

0.001	0.005	0.074	0.25	2.0	9	24	76 mm
Colloids	Clay	Silt	Fine / Coarse (Sand)		Fine	Medium	Coarse (Gravel) / Boulders

(d)

0.002	0.05	0.25	0.5	10	20	76 mm
Clay	Silt	Very fine / Fine / Medium / Coarse / Very coarse (Sand)		Fine	Medium (Gravel)	Cobbles

(e)

0.005	0.05	0.25	2.0 mm
Clay	Silt	Fine / Coarse (Sand)	Gravel

(f)

0.002	0.02	0.2	2.0 mm
Clay	Silt	Fine / Coarse (Sand)	Gravel

270	140	40	10	½ in.	3 in.
200	60	20	4	¾ in.	

Sieve Sizes

*(a) M. I. T. and British Standards Institution; (b) American Society for Testing and Materials; (c) American Association of State Highway Officials; (d) U. S. Department of Agriculture; (e) Federal Aviation Agency; (f) International Society of Soil Science.

Textural Classification

Soil texture (fabric of soil) is defined by the percentage of size group contained in the sample. Several methods have been devised for this purpose. The best-known method of textural classification was developed by the U.S. Department of Agriculture and is based on the triangular chart of Fig. 2-1, for particles smaller than 2 mm in size (Problem 2.8).

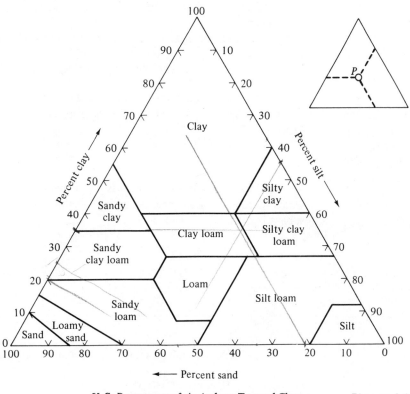

U. S. Department of Agriculture Textural Classification Chart.
(M. Peech et. al., "Methods of Soil Analysis for Soil-Fertility Investigation," U. S. Department of Agriculture, Circular 757, 1947.)

Figure 2-1

2.4 CLASSIFICATIONS BY USE

AASHO Classification System

One of the first (oldest) engineering classification systems was devised in 1929 by the Bureau of Public Roads for the evaluation of soils in highway constructions. It was revised several times since then. The latest revision, proposed in

1945 and termed the Revised American Association of State Highway Officials Classification System (AASHO Classification System), is based on the criteria of Table 2-4. In this system, soils are divided into two major categories according to their sieve analysis. They are *granular soils* (containing 35 percent or less grains passing a No. 200 mesh sieve), *clay and silt soils* (containing more than 35 percent material passing a No. 200 mesh sieve).

The soil components in this system are defined as follows:

Gravel, material passing a 3-in. sieve but retained on a No. 10 sieve.
Coarse sand, material passing a No. 10 sieve but retained on a No. 40 sieve.
Fine sand, material passing a No. 40 sieve and retained on a No. 200 sieve.
Combined silt and clay, material passing the No. 200 sieve.

The composition, characteristics, and the field performance of these different groups are described in Table 2-4. The application of this table is shown in Problem 2.9.

Group Index

The AASHO Classification System is supplemented by the group index (GI), which permits a more specific rating of the expected performance of the subgrade materials, as shown below.

Group index	0	1	2	3–4	5–9	10–20
Rating	Excellent	Very good	Good	Fair	Poor	Very poor

This index is determined either by the empirical formula (Problem 2.10) as

$$GI = 0.2a + 0.005ac + 0.01bd \qquad (2\text{-}1)$$

where a = portion of percentage passing No. 200 sieve greater than 35 percent and not exceeding 75 percent, expressed as a positive whole number (0 to 40).

b = portion of percentage passing No. 200 sieve greater than 15 percent and not exceeding 55 percent, expressed as a positive whole number (0 to 40).

c = portion of the numerical liquid limit (LL) greater than 40 and not exceeding 60, expressed as a whole number (0 to 20).

d = portion of the numerical plasticity index (PI) greater than 10 and not exceeding 30, expressed as a positive whole number (0 to 20).

$PI = LL - PL$

AASHO Classification System
(Revised USBPR Classification System)

TABLE 2-4

Group	Sub-group	PERCENT PASSING U.S. SIEVE NO. 10	40	200	CHARACTER OF FRACTION PASSING NO. 40 SIEVE Liquid Limit	Plasticity Index	Group Index No.	Soil Description	Subgrade Rating
A-1			50 max	25 max		6 max	0	Well-graded gravel or sand; may include fines.	
	A-1-a	50 max	50 max	15 max		6 max	0	Largely gravel but can include sand and fines.	
	A-1-b		50 max	25 max		6 max	0	Gravelly sand or graded sand; may include fines.	
A-2*				35 max			0-4	Sands and gravels with excessive fines.	Excellent to good
	A-2-4			35 max	40 max	10 max	0	Sands, gravels with low-plasticity silt fines.	
	A-2-5			35 max	41 min	10 max	0	Sands, gravels with elastic silt fines.	
	A-2-6			35 max	40 max	11 min	4 max	Sands, gravels with clay fines.	
	A-2-7			35 max	35 min	11 min	4 max	Sands, gravels with highly plastic clay fines.	
A-3			51 min	10 max		Nonplastic	0	Fine sands.	
A-4				36 min	40 max	10 max	8 max	Low-compressibility silts.	
A-5				36 min	41 min	10 max	12 max	High-compressibility silts, micaceous silts.	
A-6				36 min	40 max	11 min	16 max	Low-to-medium-compressibility clays.	Fair to poor
A-7				36 min	41 min	11 min	20 max	High-compressibility clays.	
	A-7-5			36 min	41 min	11 min†	20 max	High-compressibility silty clays.	
	A-7-6			36 min	41 min	11 min†	20 max	High-compressibility, high-volume-change clays.	
A-8								Peat, highly organic soils.	Unsatisfactory

*Group A-2 includes all soils having 35 percent or less passing a No. 200 sieve that cannot be classed as A-1 or A-3.
†Plasticity index of A-7–5 subgroup is equal to or less than LL-30. Plasticity index of A-7–6 subgroup is greater than LL-30.

Report of committee on Classification of Highway Subgrade Materials, Proceedings, Highway Research Board, Vol. 25, 1945, pp. 377-378.

or by the GI charts of Fig. 2-2 as the sum of both vertical readings (Problem 2.11).

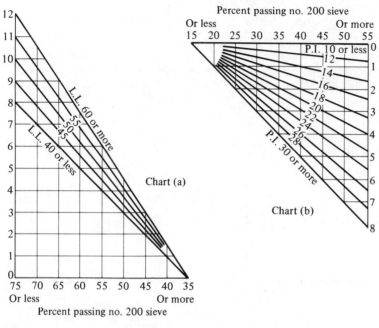

Figure 2-2   *Group index charts.*
("Report of Committee on Classification of Highway Subgrading Materials" *Proceedings, Highway Research Board*, Vol. 25, 1945, p. 375.)

Unified Soil Classification System

The Unified Soil Classification System is a product of the Airfield Classification System (AC System) developed by A. Casagrande at request of the U.S. Army Corps of Engineers in 1942. Later on the AC System was expanded and revised in cooperation with the U.S. Bureau of Reclamation, in order to apply to engineering projects other than airfields. In 1952 this system, under its present designation, was adopted by the agencies of the federal government and by state highway departments and became the most widely used classification system.

In this system (Table 2-5), the soils are classified on the basis of their action and reaction, in addition to their textural makeup. The soils are first divided into *coarse-grained, fine-grained,* and *fibrous soil.* Furthermore, the soils are designated by a group symbol consisting of a *prefix* (G = gravel, S = sand, M = silt, C = clay, O = organic clay, Pt = organic swamp soil) indicating the *soil type* and a *suffix* (W = well graded, U = uniformly graded, P = poorly graded, F = excess of fines, L = low plasticity or compressibility for clays and silts, respectively, liquid limit 20–35, I = intermediate plasticity or compressibility for clays and silts, respectively, liquid limit 35–50, H = high plasticity or compressibility for clays and silts, respectively, liquid limit > 50), indicating the *soil subtype.*

Six solid components in this system are shown below.

Cobbles:	above 3 inches	Medium sand:	No. 10 sieve to a No. 40 sieve
Gravel:	3 inches to a No. 4 sieve	Fine sand:	No. 40 sieve to a No. 200 sieve
Coarse sand:	No. 4 sieve to a No. 10 sieve	Fines:	passing a No. 200 sieve

The coarse-grained soils are grouped on the basis of the percentage of gravel, sand, fines, and the shape of the grain-size distribution curve. The coefficient of uniformity C_u and the gradation C_g are used to judge the grain-size distribution curve of coarse-grained soils. The empirical formulas for these coefficients are

$$C_u = \frac{D_{60}}{D_{10}} \qquad C_g = \frac{D_{30}^2}{D_{10}D_{60}} \qquad\qquad (2\text{-}2,\ 3)$$

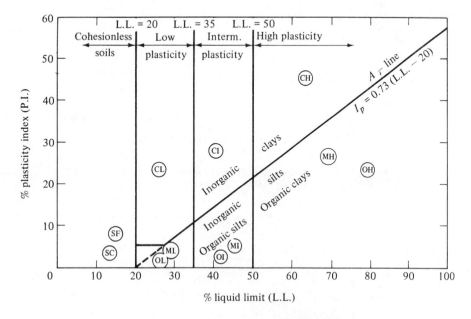

Plasticity chart. **Figure 2-3**
(A. Casagrande, "Classification and Identification of Soils," *Trans, Am. Soc. of Civ. Eng.,* Vol. 113, 1948, p. 901.)

TABLE 2-5   *Unified Soil Classification System*

Major Divisions		Symbol Letter	Name	Value as Subgrade When Not Subject to Frost Action	Value as Subbase When Not Subject to Frost Action
Coarse-grained soils	Gravel and gravelly soils	GW	Well-graded gravel-sand mixtures, little or no fines.	Excellent	Excellent
		GP	Poorly graded gravels or gravel-sand mixtures, little or no fines.	Good to excellent	Good
		GM d	Silty gravels, gravel-sand-silt mixtures.	Good to excellent	Good
		GM u		Good	Fair
		GC	Clayey gravels, gravel-sand-clay mixtures.	Good	Fair
	Sand and sandy soils	SW	Well-graded sands or gravelly sands, little or no fines.	Good	Fair to good
		SP	Poorly graded sands or gravelly sands, little or no fines.	Fair to good	Fair
		SM d	Silty sands, sand-silt mixtures.	Fair to good	Fair to good
		SM u		Fair	Poor to fair
		SC	Clayey sands, sand-clay mixtures.	Poor to fair	Poor
Fine-grained soils	Silts and clays LL is less than 50	ML	Inorganic silts and very fine sands, rock flour, silty or clayey fine sands or clayey silts with slight plasticity.	Poor to fair	Not suitable
		CL	Inorganic clays of low to medium plasticity, gravelly clays, sandy clays, silty clays, lean clays	Poor to fair	Not suitable
		OL	Organic silts and organic silt-clays of low plasticity.	Poor	Not suitable
	Silts and clays LL is greater than 50	MH	Inorganic silts, micaceous or diatomaceous fine sandy or silty soils, elastic silts.	Poor	Not suitable
		CH	Inorganic clays of high plasticity, fat clays.	Poor to fair	Not suitable
		OH	Organic clays of medium to high plasticity, organic silts.	Poor to very poor	Not suitable
Highly organic soils		Pt	Peat and other highly organic soils	Not suitable	Not suitable

Value as Base When Not Subject to Frost	Potential Frost Action	Compressibility and Expansion	Drainage Characteristics	Compaction Equipment	Unit Dry Weight lb per cu. ft.
Good	None to very slight	Almost none	Excellent	Crawler-type tractor, rubber-tired roller, steel wheeled roller	125–140
Fair to good	None to very slight	Almost none	Excellent	Crawler-type tractor, rubber-tired roller, steel wheeled roller	110–140
Fair to good	Slight to medium	Very slight	Fair to poor	Rubber-tired roller, sheepsfoot roller, close control of moisture	125–145
Poor to not suitable	Slight to medium	Slight	Poor to practically impervious	Rubber-tired roller, sheepsfoot roller	115–135
Poor to not suitable	Slight to medium	Slight	Poor to practically impervious	Rubber-tired roller, sheepsfoot roller	130–145
Poor	None to very slight	Almost none	Excellent	Crawler-type tractor, rubber-tired roller	110–130
Poor to not suitable	None to very slight	Almost none	Excellent	Crawler-type tractor, rubber-tired roller	105–135
Poor	Slight to high	Very slight	Fair to poor	Rubber-tired roller, sheepsfoot roller, close control of moisture	120–135
Not suitable	Slight to high	Slight to medium	Poor to practically impervious	Rubber-tired roller, sheepsfoot roller	100–130
Not suitable	Slight to high	Slight to medium	Poor to practically impervious	Rubber-tired roller, sheepsfoot roller	100–135
Not suitable	Medium to very high	Slight to medium	Fair to poor	Rubber-tired roller, sheepsfoot roller, close control of moisture	90–130
Not suitable	Medium to high	Medium	Practically impervious	Rubber-tired roller, sheepsfoot roller	90–130
Not suitable	Medium to high	Medium to high	Poor	Rubber-tired roller, sheepsfoot roller	90–105
Not suitable	Medium to very high	High	Fair to poor	Sheepsfoot roller, rubber-tired roller	80–105
Not suitable	Medium	High	Practically impervious	Sheepsfoot roller, rubber-tired roller	80–115
Not suitable	Medium	High	Practically impervious	Sheepsfoot roller, rubber-tired roller	80–110
Not suitable	Slight	Very high	Fair to poor	Compaction not practical	

Road Research Laboratory, DSIR, Soil Mechanics for Road Engineers, *H.M.S.O., London, 1951, p. 67.*

where D_{10}, D_{30}, D_{60} are the grain-size diameters at 10, 30, and 60 percent, respectively, on the grain-size distribution curve. This curve is a graphical representation of grain-size distribution in a soil sample and is plotted arithmetically or logarithmetically (Problem 2.12).

The fine-grained soils are characterized by the plasticity index (PI) incorporated in the classification by use of the plasticity chart of Fig. 2-3, in which the abcissa is the liquid limit (LL) and the ordinate gives the plasticity index (PI). A straight line (A) drawn diagonally across the chart, given empirically as

$$PI = 0.73(LL - 20) \tag{2-4}$$

divides the chart into six basic areas. The group to which the soil belongs is determined by the designation of the area that contains the point representing the values of PI and LL for the soil. In general, clay (C) plots above the A line and silts (M) plot below the A line. The silts and clays are further divided on the basis of low (L) or high (H) liquid limit (Problems 2.13 and 2.15).

The description of soil groups is based on visual and manual procedures. Applicable observations and tests related to the material in place are recommended as an integral part of the system. Soil groups are rated according to their engineering performance and characteristics, like frost susceptibility and swelling. Table 2-5 shows the correlations between the different groups and the associated characteristics.

SOLVED PROBLEMS

Geological and Pedological Classifications

2.1   Define residual soils and describe the major soil groups of these soils. Also, give examples of each.

Answer   Residual soils are found in places where products of rock weathering are not transported as sediments but accumulate in place and where the rate of rock decomposition exceeds the rate of removal of the products of decomposition.

When mature, the residual soil profile is divided into three zones:

1.   The upper zone, where there is a high degree of leaching and removal of material.
2.   The intermediate zone, where there is some weathering at the top but some deposition toward the bottom.
3.   The partially weathered zone this is the transition zone between weathered material to the unweathered parent rock.

The three major divisions of residual soils, their descriptions, and their subdivisions on the basis of the combined effect of climate, vegetation, and topography are given in Table P-2.1

2.2   Classify transported soils according to their origin, mode of formation, and unit landforms.

Order	Suborders	Great Soil Groups
Zonal Soils Mature soils characterized by well-differentiated horizons and profiles that differ noticeably according to the climatic zone in which they occur. They are found in great areas where the land is well drained but not too steep.	1. Soils of the cold zone 2. Light-colored soils of arid regions	Tundra soils Desert soils Red desert soils Sierozem Brown soils Reddish-brown soils
	3. Dark-colored soils of semiarid, subhumid, and humid grassland	Chestnut soils Reddish chestnut soils Chernozem soils Prairie soils Reddish prairie soils
	4. Soils of the forest-grassland transition	Degraded chernozem Noncalcic brown or shantung brown soils
	5. Light-colored podzolized soils of the timbered regions	Podzol soils Gray wooded or gray podzolic soils Brown podzolic soils Gray-brown podzolic soils Red-yellow podzolic soils
	6. Lateritic soils of forested warm-temperate and tropical regions	Reddish-brown lateritic soils Yellowish-brown lateritic soils Laterite soils
Intrazonal Soils Soils with well-developed characteristics resulting from some influential local factor of relief or parent rock. They are usually local in occurrence. Bog soils, peats, and salt soils are typical examples.	1. Halomorphic (saline and alkali) soils of imperfectly drained arid regions and littoral deposits	Solonchak or saline soils Solonetz soils Soloth soils
	2. Hydromorphic soils of marshes, swamps, seep areas, and flats	Humic-glei soils (includes wiesenboden) Alpine meadow soils Bog soils Half-bog soils Low-humic-glei soils Planosols Groundwater poudzol soils Groundwater laterite soils
	3. Calcimorphic soils	Brown forest soils (braunerde) Rendzina soils
Azonal Soils These soils are relatively young and reflect to a minimum degree the effects of environment. They do not have profile development and structure developed from the soil-forming processes. Alluvial soils of flood plains and dry sands along large lakes are examples.		Lithosols Regosols (includes dry sands) Alluvial soils

Answer Soils can be transported by any of the four agents listed in Section 2.2, page 26. Transportation affects sediments in two major ways:

1. It alters particle shape, size, and texture by abrasion, grinding, impact, and solution;
2. It sorts the particles.

Table P-2.2 shows the classification of transported soils according to origin. Most of the soil units (landforms) appearing in the table have distinct physical and engineering characteristics.

TABLE P-2.2

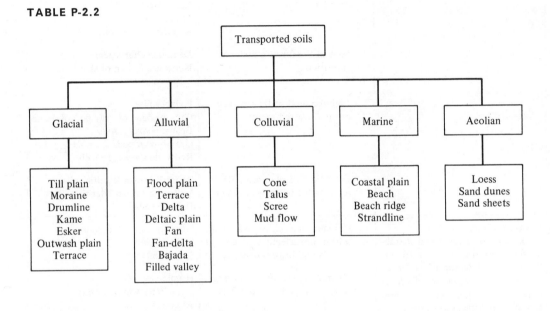

2.3 Define the soil series in the pedologic system of classification.

Answer Soils within each great soil group (Table P-2.1) are divided into series. A soil series is a number of soils within each great soil group that have uniform development (the same age, climate, vegetation, and relief) and similar parent material. The only difference is that the texture of the surface soil or the horizon may vary in the profiles of the same soil series.

The soil series is further subdivided into the final pedologic classification unit, called the soil type. The soil type is made up of the soil series name plus the textural classification of the (A) horizon.

For example, if the textures of the (A) horizon of a soil series named Miami are classified texturally (Section 2.3) as sand and sandy loam, the soil type in each case would be Miami sand and Miami sandy loam. Both soils will have the same (B) and (C) horizons (parent material) and would have been found under the same conditions of climate, vegetation, and topography.

2.4 Discuss the value of the pedologic system of soil classification in engineering applications.

The major advantages in the use of the pedologic classification are   *Answer*

1. The soil series, as a unit in the classification system, reflects much information concerning the soil for which there is no established testing or measuring procedure (e.g., susceptibility of soil to erosion by wind or water, its internal and external drainage characteristics, the depth to the water table during various seasons of the year).
2. Field profiling can be reduced to only check boring. Detailed profiling would be confined to areas of transition from one soil to another. This is in contrast to other classification methods (Section 2.4) where the sample taken for classification only represents the immediate area from which it was obtained.
3. The system is natural and based on more scientific basis than other methods of classification, which makes it open to expansion for new information regarding soils.
4. It is more adequate as a tool for the exchange of information regarding soil types in different parts of the world.

2.5 List the disadvantages of the pedologic classification of soils in engineering.

The main disadvantages of the system are   *Answer*

1. Proper interpretation and integration have not been done for deeper strata.
2. The significance of the method is also a function of the type of deposit or formation. In areas where bedrock is exposed or where very shallow soils exist on rough terrain, the value of the method vanishes.
3. The division indicated as important to agronomists may exist in the upper 12 inches, which is referred to by engineers as a "topsoil" and is normally wasted.

Grain-Size Classification

2.6 Three soils are represented by their particle-size distribution curves *A*, *B*, *C* in Fig. P-2.6. Classify these soils according to the standards of the International Society of Soil Science.

Using the standard of the International Society of Soil Science (Table 2.3, Section   *Solution*
2.3), the percentages of each size group for soils *A*, *B*, and *C* are

Soil Type	*A*	*B*	*C*
Percent clay	10	3	20
Percent silt	30	40	55
Percent sand	60	57	25

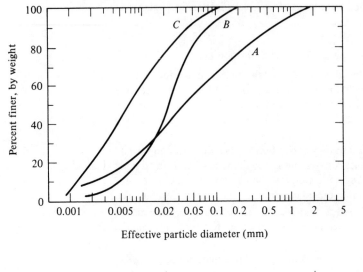

Effective particle diameter (mm)

| Clay | | Silt | | Sand | | Gravel (Int. Soc.) |
| Clay | | Silt | | Sand | | Gravel (U.S. Dept. of Agriculture) |

Figure P-2.6

2.7 Classify the soils of Fig. P-2.6 according to the standards of the U.S. Department of Agriculture.

Solution Using the U.S. Department of Agriculture Standards (Table 2.3), the percentages of each size group for soils *A,B,* and *C* are as follows:

Soil Type	*A*	*B*	*C*
Percent clay	10	3	20
Percent silt	45	77	70
Percent sand	45	20	10

Textural Classification

2.8 An inorganic soil sample has the following grain-size characteristics:

Size (mm)	2	0.074	0.050	0.005	0.002
Percent passing	100	72	68	30	20

Classify this soil according to the U.S. Department of Agriculture soil-texture classification system.

First determine the percentage of each soil group (clay, silt, and sand) in this soil.   *Answer* Next, using the U.S. Department of Agriculture size-classification scale (Table 2-3), list the percentages.

$$\text{Clay size} = 20 \text{ percent}$$

$$\text{Silt size } (68\text{--}20) = 48 \text{ percent}$$

$$\text{Sand size } (100\text{--}68) = 32 \text{ percent}$$

Then using the triangular chart (Fig. 2-1), plot the three percentages of clay, silt, and sand on the three sides of the triangle, and draw three lines extending from these points to the inside of the chart, as shown in the figure. These three lines will intersect at point P, which is located in the area designated as "loam." Thus the texture class of this soil is loam.

Classification by Use

2.9   The soil represented by curve C in Fig. P-2.6 was found to be nonplastic (i.e., a plastic limit (PL) test could not be performed), and its liquid limit (LL) is 45 percent. Classify the soil according to the AASHO System.

Percent passing No. 200 sieve for soil represented by grain-size curve C in Fig. P-2.6   *Answer* is 94 percent (No. 200 sieve opening size $= 0.074$ mm). Since this is more than 35 percent, the soil belongs to one of the fine-grain groups (A-4, A-5, A-6, or A-7).
    For additional specific identification, the LL and PI are used.

LL $= 45\,\%$ ($>41\,\%$), (PL) $= 0$ (soil is nonplastic)

PI $= 0$ (soil is nonplastic and PI $< 11\,\%$)

Referring to Table 2-4, the soil is classified as A-5.

2.10   For the same soil in Problem 2.9, compute the group index (GI) of the soil, using the empirical formula (2-1).

Solution

$a = 75 - 35 = 40$ (75 is substituted for 94, since the range of a is from 35 to 75)

$b = 55 - 15 = 40$ (55 is substituted for 94, since the range of b is from 15 to 55)

$c = 45 - 40 = 5$

$d = 0$ (soil is nonplastic)

Substituting these values in (2-1) yields

$$\text{GI} = (0.2)(40) + (0.005)(40 \times 5) + (0.01 \times 40 \times 0) = 9.0$$

2.11   Compute the group index for the soil of Problem 2.9, using the GI charts of Fig. 2-2. Show how the final classification of this soil is expressed according to the AASHO System.

Answer   The group index (GI) is the sum of the two readings on the vertical scale of charts (a) and (b) of Fig. 2-2. With values of LL, PI, and percent passing No. 200 sieve calculated in Problem 2.10,

$$GI = 9.0 + 0 = 9.0$$

The group index is given in parentheses after the soil group number. In this case, the soil is classified according to the AASHO System as A-5 (9).

2.12   Three soils were sieved and the results plotted on a semilog paper, as shown in Fig. P-2.12. From the shape of the curves, describe the size distribution of each sample. Also, compute the uniformity coefficient C_u and the gradation coefficient C_g, and comment on the results.

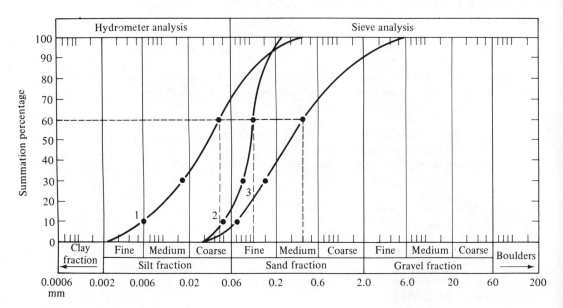

Figure P-2.12

Solution   A uniform soil, where all the particles are approximately the same size, has an almost vertical size distribution curve. A well-graded soil, containing a wide range of particle size, has a curve spread evenly across the chart.

From the shape of these three curves, sample No. 1 is well-graded sandy silt; sample No. 2 is uniform fine sand; and sample No. 3 is well-graded sand.

Curve 1

$$C_u = \frac{0.045}{0.006} = 7.50 \qquad C_g = \frac{(0.016)^2}{(0.006)(0.045)} = 0.95$$

Curve 2

$$C_u = \frac{0.12}{0.05} = 2.40 \qquad C_g = \frac{(0.082)^2}{(0.05)(0.12)} = 1.12$$

Curve 3

$$C_u = \frac{0.40}{0.07} = 5.72 \qquad C_g = \frac{(0.15)^2}{(0.07)(0.40)} = 0.81$$

A uniform soil will have a C_u value approaching 1.0, whereas a well-graded soil will have a higher value of C_u.

2.13 A soil sample was found to have 55 percent of effective particle diameter less than 0.074 mm, LL = 60, and PI = 35. Classify this soil according to the Unified System.

Since more than 50 percent of the material is finer than 0.074 mm (No. 200 sieve size) and the liquid limit is greater than 50 percent, it belongs to the group of clays and silts of high compressibility in the Unified System. On the plasticity chart (Fig. 2-3), it plots above the A line in the zone designated as "CH" and is therefore a clay with high plasticity, CH.   *Answer*

2.14 List the advantages and disadvantages of the AASHO System of soil classification described in Section 2.4 and depicted by Table 2-4.

The main advantages and disadvantages of the AASHO Soil Classification System are summarized in Table P-2.14.   *Answer*

TABLE P-2.14

Main Advantages of the AASHO Soil Classification	Main Disadvantages of the AASHO Soil Classification
1. Tests used in the procedure of classification are widely known among engineers. The system therefore provides an excellent medium for translating engineering information and experience between highway organizations. 2. From a practical standpoint, the system is easy to apply, and the group classification and the group index could always be given in addition to any other tests. 3. The number of complicated and expensive additional tests, if needed, can be performed on typical samples rather than on all samples from individual projects.	1. No place for organic materials in the system. 2. Possibility of overlapping between soil properties of two groups. It is difficult to have a sharp borderline between soils. 3. The symbols used in the system are not descriptive. Most highway engineers, however, have a clear meaning of the symbols as far as engineering performance of soils is concerned.

2.15 List the main advantages and disadvantages of the Unified Soil Classification System described in Section 2.4 and depicted by Table 2-5.

Answer The main advantages and disadvantages of the Unified Soil Classification System are summarized in Table P-2.15.

TABLE P-2.15

Main Advantages of the Unified Soil Classification System	Main Disadvantages of the Unified Soil Classification System
1. It is more flexible and thus suitable for classification of a wider variety of soils than the AASHO System. 2. The symbols used are descriptive and easy to associate with engineering properties of soils. 3. Certain engineering properties can be assigned to the soil groups given in the system; therefore correlations with design procedures can be made more easily.	1. The No. 10 sieve in the AASHO System is better as an upper limit for sand than the No. 4 sieve in this system. The No. 10 sieve as an upper limit agrees with the accepted sieve separation in concrete and highway base-course technology. 2. The 50 percent passing a No. 200 sieve as a limit for distinguishing fine-grained and coarse-grained soils is rather high. It may be that with less than 50 percent passing a No. 200 sieve would make the material fall in the fine group.

SUPPLEMENTARY PROBLEMS

Geological and Pedological Classification

2.16 Discuss the effect of each of the five genetic soil-forming factors on the development of a soil profile.

Answer For a detailed discussion on the influence of the soil-forming factors on the development of a soil profile, refer to standard texts of geomorphology, geology, and engineering geology. Only a brief outline will be attempted in the following paragraphs.

1. *Parent material*: Soils owe some of their caracteristics to the kind of parent materials from which they were derived. Thus residual soils derived from limestone, sandstone, granite, etc. will differ somewhat in their chacteristics.

2. *Climate*: Temperature and rainfall are the significant climatic factors. Warm and humid climates encourage rapid weathering of the minerals of the parent rock. Rainfall intensity affects the leaching of certain minerals from the upper parts of the soil profile and depositing it in the lower zones.

3. *Topography*: It affects the percolation and infiltration of water through the weathered soil profile. It influences the rate of erosion at the surface and therefore the depth of the exposed soil profile.

4. *Vegetation*: The amount and type of products resulting from decay of vegetation, such as organic acids, are significant factors in weathering.

5. *Time*: time is essential in allowing weathering processes to take place and the soil profile to develop. A long time allows leaching of minerals from rainfall and other weathering processes to take place.

2.17 List the steps followed in making a detailed engineering soil map, using the pedological system of classification.

Answer

Preparation of Detailed Engineering   **TABLE P-2.17**
Soil Maps

1. A careful review of all available information (such as aerial photographs, surface geology maps, U.S. Dept. of Agriculture county soil maps) is made.
2. The engineer usually makes a preliminary study of road and railroad cuts, erosion, soil exposures of agriculture use and vegetation, drainage patterns, and mining operations, including clay and gravel pits.
3. After a review of this information, the soil engineer proceeds to the field to make the detailed soil map.
4. The final classification and location of boundaries are determined by field examination of auger borings and detailed soil profiles.
5. The soils are then grouped, on the basis of their characteristics, both internal and external, into mapping units.

Grain-Size Classification

2.18 The following results were obtained from a sieve analysis of a soil sample:

Sieve size	4	10	20	40	100	200	270
Percent passing	95	89.1	74.8	58.2	40.1	3.8	1.2

Classify the soil according to the MIT (British Standards Institution) grain-size classification system (Table 2-3).

12 percent gravel, 86 percent sand, 2 percent silt.   *Answer*

Texture Classification

2.19 An inorganic soil was found to have the following size distribution: clay size, 16 percent; silt size, 56 percent; sand size, 28 percent. Classify this soil according to the U.S. Department of Agriculture textural system.

Silty loam.   *Answer*

Classification by Use

2.20 What is the major deficiency of the AASHO and the Unified systems of soil classifications?

Answer Both systems deal with remolded soil, and the structure of soil in place is given little or no consideration.

2.21 Classify the soil samples defined below according to the Unified and AASHO classification systems and compute the group index (GI) for each sample.

Samples	1	2	3	4	5	6
Sieve Analysis			Percentage Passing			
No. 4 sieve	97	100	98	93	85	100
No. 10 sieve	96	100	94	87	80	93
No. 40 sieve	93	94	80	68	60	69
No. 200 sieve	87	68	57	46	28	32
No. 270 sieve	84	63	50	47	27	26
0.005 mm	50	21	20	16	9	9
0.001 mm	25	10	15	8	3	3
Index Properties						
Liquid Limit	32	26	47	31	21	42
Plastic Limit	23	15	35	2	17	34

Answer CL, A-4 (8); CL, A-6 (7); ML, A-7-5 (6); SC, A-4 (2); SC, A-2-4 (0); SM, A-2-5 (0).

rheology

3

3.1 INTRODUCTION

Concept

The study of time-dependent deformations of materials is called the *rheology* (from the Greek word *rheos*, moving flow) and, in this particular case, the designation of soil rheology is used. Since the engineering soils exhibit elastic, plastic, elastoplastic, viscoelastic, and viscoplastic characteristics, their rheological behaviors are more complex than those of other construction materials.

The basis of rheological study is the development of the constitutive equations, stating the cause-effect (load-displacement) relationships. Because the causes and effects are functions of time and the soil material constants are functions of time and moisture content, the constitutive equations are frequently transcendent equations and lead to a nonlinear analysis.

Rheological Models

For the analysis, the real (physical) soil system is replaced by an *ideal mechanical model*, called the rheological model, which is a composition of basic units of springs and dashpots in series or in parallels.

A *single spring element* known as the Hookean model represents a linear cause-effect relationship, and its performance is independent of time (purely elastic phenomenon).

A *single dashpot element* known as the Newtonian model represents a nonlinear cause-effect relationship, and its performance is dependent on time (purely viscous phenomenon).

These two rheological models are called the *elementary models*; their combinations required by the soil behavior are then termed the *composite models* and are frequently designated by the names of their originators as the Kelvin, the Maxwell, the Burger, the Bingham models, and so on. Their development, mathematical treatment, and simple applications are shown in this chapter; their more extensive applications and refinements are introduced in Chapter 6.

3.2 ELEMENTARY MODELS

Classification

Three basic rheological models are of practical interest: they are the *Hookean model* (perfect elastic model), the Newtonian model (perfect viscous model), and the yield stress model (perfect strain lack model). All three models characterize the stress-strain relationship (constitutive law) in terms of the material constants known from experiments.

Hookean Model

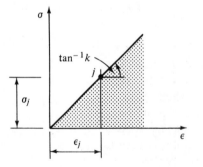

(a)   Stress-strain diagram

If the stress σ is a linear function of the strain ϵ (Fig. 3-1a), the stress-strain relationship is

$$\sigma = k\epsilon \tag{3-1}$$

where k is the constant of proportionality, also known as the modulus of elasticity (lb/in². or kg/cm²).

The model representation of (3-1) is shown in Fig. 3-1b, where k is the spring constant.

Newtonian Model

If the stress σ is a linear function of the rate of change in strain ϵ with respect to time t (Fig. 3-2a), the stress-strain relationship (Problem 3.8) is

$$\sigma = \frac{\eta}{t}(\epsilon + \epsilon_0) \tag{3-2}$$

where η is the constant of viscosity (lb/in² $\times$ t, kg/cm² $\times$ t) and ϵ_0 is the strain at $t = 0$.

The model representation of (3-2) is shown in Fig. 3-2b, where η is the dashpot constant.

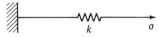

(b)   Model

Figure 3-1   *Hookean model*

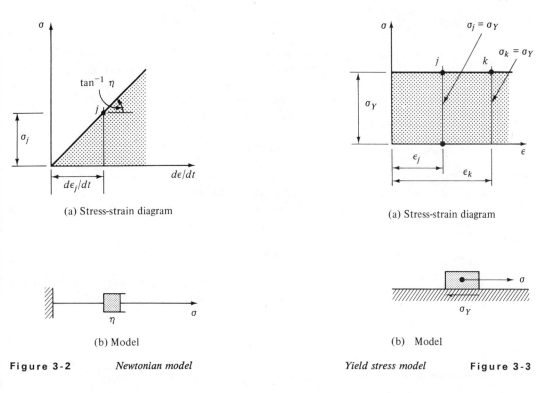

(a) Stress-strain diagram

(a) Stress-strain diagram

(b) Model

(b) Model

Figure 3-2   *Newtonian model*

Yield stress model   **Figure 3-3**

Yield Stress Model

In this case, the stress σ generates a strain ϵ (Fig. 3-3a) if and only if

$$\sigma \geq \sigma_Y \tag{3-3}$$

where σ_Y represents the minimum stress under which no strain occurs, called the yield stress.

The model representation of (3-3) is shown in Fig. 3-3b, where σ_Y is the friction resistance.

3.3 SIMPLE COMPOSITE MODELS

Construction

Since the stress-strain relationships of most soils do not follow the load-displacement pattern of the elementary models, the simple composite models constructed as a combination of two elementary models must be introduced for their representation. They are the *St. Venant model*, the *Kelvin model*, and the *Maxwell model*.

St. Venant Model

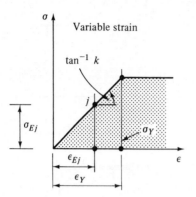

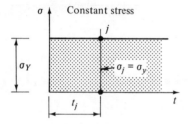

(a) Stress-strain diagram

(b) Stress-time diagram

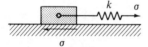

(c) Model

Figure 3-4   *St. Venant model*

The combination of (3-1) and (3-3) produces an elasto-plastic response (Fig. 3-4a and b), given by

$$\sigma_Y \geq \sigma_E = k\epsilon_E$$
$$\sigma_Y \leq k\epsilon_Y \tag{3-4}$$

where σ_E, ϵ_E are the stress and strain in the elastic range, respectively, σ_Y, ϵ_Y are the stress and strain at the yield point, respectively, and k is the constant of (3-1).

The model representation of (3-4) is shown in Fig. 3-4c, where σ_Y is the friction resistance and k is the spring constant.

Kelvin Model

The parallel combination of (3-1) and (3-2) produces a viscoelastic response (Fig. 3-5a, b), given by

$$\sigma = k\epsilon + \eta \frac{d\epsilon}{dt} \tag{3-5a}$$

the solution of which, for $\sigma_{t=0} = \sigma_{t=t} = \sigma_0$, $\epsilon_{t=0} = \epsilon_0$ (Problem 3.10), is

$$\epsilon = \frac{\sigma_0}{k}(1 - e^{-\alpha t}) + \epsilon_0 e^{-\alpha t} \tag{3-5b}$$

where $e = 2.71828\ldots =$ base of natural logarithm, σ_0, ϵ_0 are the stress and strain at $t = 0$, $\alpha = k/\eta$, and η, k are the constants of (3-1) and (3-2) respectively.

The model representation of (3-5) is shown in Fig. 3-5c, where k is again the spring constant and η is the dashpot constant. This model is sometimes called the *Voigt model*.

Maxwell Model

The series combination of (3-1) and (3-2) produces an inverse viscoelastic response (Fig. 3-6a and b), given by

$$\frac{d\sigma}{dt} + \frac{k}{\eta}\sigma = k\frac{d\epsilon}{dt} \tag{3-6a}$$

the solution of which, for $\sigma_{t=0} = \sigma_0$ (Problem 3.11), is

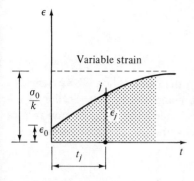

(a) Strain-time diagram

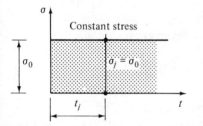

(b) Stress-time diagram

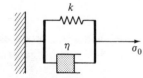

(c) Model

Figure 3-5   *Kelvin model*

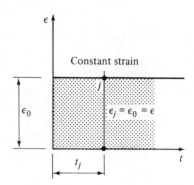

(a) Strain-time diagram

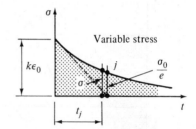

(b) Stress-time diagram

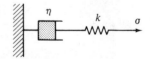

(c) Model

Maxwell model   **Figure 3-6**

$$\sigma = e^{-\alpha t}\left(k \int \frac{d\epsilon}{dt} e^{\alpha t}\, dt + \sigma_0\right) \qquad\qquad (3\text{-}6b)$$

where the symbols have the same meaning as in (3-5).

The model representation of (3-6) is shown in Fig. 3-5c, where k and η are the constants of (3-1) and (3-2) respectively. It should be noted that the diagrams of Fig. 3-6a and b are plotted for the special case

$$\epsilon = \epsilon_0 \qquad \frac{d\epsilon}{dt} = 0.$$

3.4 *COMPLEX COMPOSITE MODELS*

Construction

For soil systems with more complex behavior, the combination of the elementary models (Section 3.2) and of the simple composite models (Section 3.3) produces a new class of rheological models, called the *complex composite models*. The most typical cases in this group are the *Bingham model*, the *standard linear model*, and the *Burger model*.

Bingham Model

The series combination of (3-2), (3-3), and (3-1) produces an elastoplastic response if $\sigma \leq \sigma_Y$ and a delayed viscoelastic response if $\sigma \geq \sigma_Y$ (Fig. 3-7a and b). The latter is governed by a modified equation (3-6), which, for $\sigma_{t=0} = \sigma_0 = \sigma_Y$ (Problem 3.12), becomes

$$\sigma = e^{-\alpha t}\left(k \int \frac{d\epsilon}{dt} e^{\alpha t}\, dt + \sigma_0\right) + \sigma_Y \tag{3-7}$$

and its symbols are those of (3-4) and (3-5).

The model representation of (3-7) is shown in Fig. 3-7c, where σ_Y is again the friction and $\epsilon = \epsilon_0$ is a constant.

Standard Linear Model

The parallel combination of (3-6) and (3-1) produces a complex viscoelastic response (Fig. 3-8a and b), governed by

$$\sigma = k_2 \epsilon + \eta \frac{d\epsilon}{dt} - \frac{\eta}{k_1} \frac{d\sigma}{dt}$$

the solution of which, for $\sigma_{t=t} = \sigma_{t=0} = \sigma_0$ (constant stress) (Problem 3.13), is

$$\epsilon = \frac{\sigma_0}{k_2}(1 - \beta\eta e^{-\beta k_2 t}) \tag{3-8}$$

In addition to the conventional symbols used before,

$$\beta = \frac{k_1}{(k_1 + k_2)\eta}$$

and k_1, k_2 are the respective spring constants.

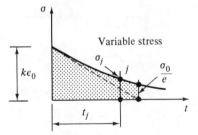

(a) Strain-time diagram

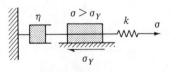

(b) Stress-time diagram

(c) Model

Figure 3-7 *Bingham model*

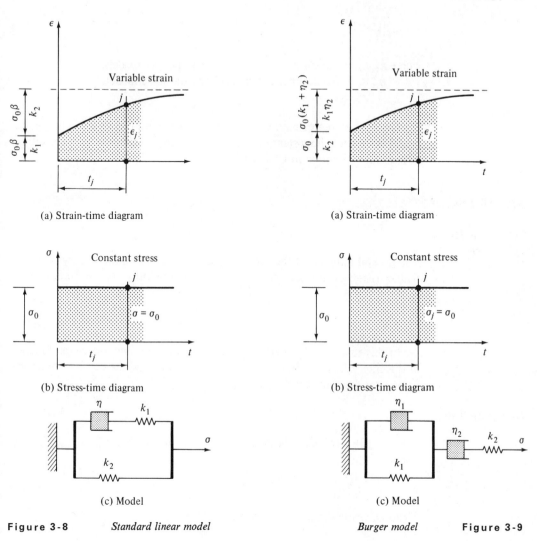

(a) Strain-time diagram

(b) Stress-time diagram

(c) Model

Figure 3-8 *Standard linear model*

(a) Strain-time diagram

(b) Stress-time diagram

(c) Model

Burger model **Figure 3-9**

The model representation of (3-8) is shown in Fig. 3-8c, where the end blocks are assumed to move parallel so that the top strain ϵ_T and the bottom strain ϵ_B remain equal during the time-dependent process $(\epsilon_{T,t=t} = \epsilon_{B,t=t})$.

Burger Model

The parallel combination of (3-5) and (3-6) again produces a complex visco-elastic response (Fig. 3-9a and b) equivalent to the consolidation response of soils. The governing equation, obtained as a superposition of (3-5) and (3-6), for $\sigma_{t=0} = \sigma_{t=t} = \sigma_0$ (Problem 3.14), is

$$\epsilon = \sigma_0 \left[\frac{1}{k_2} + \frac{t}{\eta_2} + \frac{1}{k_1} (1 - e^{-\alpha_2 t}) \right] \tag{3-9}$$

where $k_1, k_2 =$ the spring constants of the respective models,

$\eta_1, \eta_2 =$ the dashpot constants of the the same models,

$\alpha_1 = k_1/\eta_1$

The model representation of (3-9) is shown in Fig. 3-9c. The Burger model is the most important simulation of the soil mechanism and is a subject of particular attention in Chapter 6.

3.5   TRANSPORT MATRICES

General Integrals

The general integrals of the preceding sections recording the history of a particular rheological model can be written symbolically for the time interval $t_i < t < t_j$ as

$$y_j = \Phi_{ji} + \psi_{ji} y_i \tag{3-10}$$

In this equation, Φ_{ji} and ψ_{ji} are forcing function and the shape function in this interval, respectively, and y_j, y_i are the values of the dependent variable (σ or ϵ) at the time stations j, i respectively.

Transport Matrix

Since (3-10) gives the relationship of two values separated by the given time interval

$$\tau_{ji} = t_j - t_i \tag{3-11}$$

it can be written in a matrix form as

$$\underbrace{\begin{bmatrix} +1 \\ y_j \end{bmatrix}}_{\hat{Y}_j} + \underbrace{\begin{bmatrix} +1 & 0 \\ \Phi_{ji} & \psi_{ji} \end{bmatrix}}_{\hat{T}_{ji}} \underbrace{\begin{bmatrix} +1 \\ y_i \end{bmatrix}}_{\hat{Y}_i} \tag{3-12}$$

where $\hat{Y}_j$, $\hat{Y}_i$ are the state vectors at j, i, respectively, and $\hat{T}_{ji}$ is the time transport matrix, the coefficients of which are constants for the given interval τ_{ji} (Problems 3.15, 3.16, and 3.17).

The upper equality $+1 = +1$ of this matrix equation is a transformation identity necessary in the development of the transport chain introduced in the subsequent article.

Once the general form of Φ_{ji} and ψ_{ji} is available for a given model, the anticipated life span of the model can be subdivided into a number of time intervals selected to correspond to the nature of the model and its load history.

If these intervals are, for example, τ_{lk}, τ_{kj}, τ_{ji}, τ_{jh}, the state vectors at the respective time stations l, k, j, i, become, respectively,

$$\hat{Y}_l = \hat{T}_{lk}\,\hat{Y}_k, \quad \hat{Y}_k = \hat{T}_{kj}\,\hat{Y}_j, \quad \hat{Y}_j = \hat{T}_{ji}\,\hat{Y}_i, \quad \hat{Y}_i = \hat{T}_{ih}\,\hat{Y}_h \qquad (3\text{-}13)$$

Then by successive substitution,

$$\begin{aligned}
\hat{Y}_j &= \hat{T}_{ji}\hat{T}_{ij}\hat{Y}_h = \hat{T}_{jh}\hat{Y}_h \\
\hat{Y}_k &= \hat{T}_{kj}\hat{T}_{ji}\hat{T}_{ih}\hat{Y}_h = \hat{T}_{kh}\hat{Y}_h \\
\hat{Y}_l &= \underbrace{\hat{T}_{lk}\hat{T}_{kj}\hat{T}_{ji}\hat{T}_{ih}}_{\hat{T}_{lh}}\hat{Y}_h = \hat{T}_{lh}\hat{Y}_h
\end{aligned} \qquad (3\text{-}14)$$

where the product of T-matrices in each equation is a new transport matrix (Problems 3.18 and 3.19).

Since the new matrix is a result of a chain product, it is designated as a transport matrix chain.

This transport matrix chain is characteristic for the given time interval, $\tau_{lh} = t_l - t_h$, and is independent of the end conditions. It transports the effect of one state vector from one time station to another time station, and it may include any number of subintervals.

Because of the recurrent nature, the transport matrix formulations of rheological problems are very suitable for computer programming as well as numerical calculations in general.

General Integral Inverse

If the *inverse relationship* is desired then (3-10) may be written as

$$y_i = \Phi_{ij} + \psi_{ij}y_j \qquad (3\text{-}15)$$

where

$$\Phi_{ij} = \frac{-\Phi_{ji}}{\psi_{ji}} \quad \text{and} \quad \psi_{ij} = \frac{1}{\psi_{ji}}$$

Transport Matrix Inverse

With (3-15) known, the inverse relationship (3-12) can be written as

$$\underbrace{\begin{bmatrix} +1 \\ y_i \end{bmatrix}}_{\hat{Y}_i} = \underbrace{\begin{bmatrix} +1 & 0 \\ \Phi_{ij} & \psi_{ij} \end{bmatrix}}_{\hat{T}_{ij}} \underbrace{\begin{bmatrix} +1 \\ y_j \end{bmatrix}}_{\hat{Y}_j} \qquad (3\text{-}16)$$

where $\hat{Y}_i$, $\hat{Y}_j$ are the same state vectors as in (3-12) but $\hat{T}_{ij}$ is a new transport matrix giving the inverse transport (from j to i).

Because

$$\hat{T}_{ij}\hat{T}_{ji} = I \qquad (3\text{-}17)$$

and since I is a unit matrix, $\hat{T}_{ij}$ is the inverse of $\hat{T}_{ji}$ and vice versa.

Transport Chain Inverse

If (3-13) is written in the inverse order as

$$\hat{Y}_h = \hat{T}_{hi}\hat{Y}_i, \quad \hat{Y}_i = \hat{T}_{ij}\hat{Y}_j, \quad \hat{Y}_j = \hat{T}_{jk}\hat{Y}_k, \quad \hat{Y}_k = \hat{T}_{kl}\hat{Y}_l \qquad (3\text{-}18)$$

the successive substitution, as in (3-14), yields

$$\begin{aligned}
\hat{Y}_j &= \hat{T}_{jk}\hat{T}_{kl}\hat{Y}_l = \hat{T}_{jl}\hat{Y}_l \\
\hat{Y}_i &= \hat{T}_{ij}\hat{T}_{jk}\hat{T}_{kl}\hat{Y}_l = \hat{T}_{il}\hat{Y}_l \\
\hat{Y}_h &= \underbrace{\hat{T}_{hi}\hat{T}_{ij}\hat{T}_{jk}\hat{T}_{kl}}_{\hat{Y}_{hl}}\hat{Y}_l = \hat{T}_{hl}\hat{Y}_l
\end{aligned} \qquad (3\text{-}19)$$

where the product of T-matrices is again a new transport matrix.

Since

$$\hat{Y}_l = \hat{T}_{lh}\hat{Y}_h \quad \text{and} \quad \hat{Y}_h = \hat{T}_{hl}\hat{Y}_l \qquad (3\text{-}20)$$

the transport chain $\hat{T}_{hl}$ is the inverse of $\hat{T}_{lh}$ and vice versa.

SOLVED PROBLEMS

General Integrals

3.1   Derive the general solution (integral) of the differential equation

$$\dot{y} + P(t)y = Q(t) \qquad (a)$$

which is linear in y and $\dot{y} = dy/dt$, and in which $P(t), Q(t)$ are continuous functions of t in a given interval (which can be infinite).

Solution   Such an equation is called a nonhomogeneous linear differential equation of the first order with variable coefficients. The integral of this equation can be conveniently obtained by means of the integrating factor

$$\mu = e^{\int P(t)\,dt}$$ (b)

Upon multiplying the given differential equation by equation (b), its new form becomes

$$\underbrace{\dot{y}e^{\int P(t)\,dt} + P(t)\,ye^{\int P(t)\,dt}}_{L} = \underbrace{Q(t)e^{\int P(t)\,dx}}_{R}$$ (c)

where the left side

$$L = \frac{d}{dt}(ye^{\int P(t)\,dt})$$ (d)

Thus the integration of $L = R$ with respect to t gives

$$ye^{\int P(t)\,dt} = \int Q(t)e^{\int P(t)\,dt}\,dt + C$$ (e)

or simply

$$y = e^{-\int P(t)\,dt}\int Q(t)e^{\int P(t)\,dt}\,dt + Ce^{-\int P(t)\,dt}$$ (f)

which is the desired general solution of (a).

In this solution, C is the constant of integration to be determined from the initial conditions of the model's history as shown in Problems 3.4 and 3.5.

3.2 Find the general solution (integral) of the differential equation

$$\dot{y} + \alpha y = \beta$$ (a)

which is linear in y and $\dot{y} = dy/dt$, and in which α and β are constants. Also, prepare a graph of this solution.

Such an equation is called a nonhomogeneous linear differential equation with *Solution* constant coefficients. The solution of this equation is found by (f) of Problem 3.1. With

$$P(t) = \alpha \qquad Q(t) = \beta$$ (b)

the general integral becomes

$$y = e^{-\int \alpha\,dt}\int \beta e^{\alpha \int\,dt}\,dt + Ce^{-\int \alpha\,dt}$$ (c)

After the indicated integrations are performed, (c) reduces to

$$y = \frac{\beta}{\alpha} + Ce^{-\alpha t}$$ (d)

where α, β are the known constant coefficients of the given differential equation (a) and C is again the constant of integration.

The shape of the curve representing (d) depends on the values of α, β, and C, which obviously leads to a large variety of graphs.

Two most typical cases are the relaxation curve ($\alpha > 0, \beta > 0, C > 0, t > 0$) and the creep curve ($\alpha > 0, \beta > 0, C < 0, t > 0$), shown in Fig. P-3.2a and b respectively.

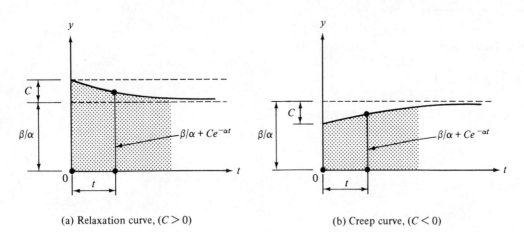

(a) Relaxation curve, ($C > 0$)　　　　　　　　(b) Creep curve, ($C < 0$)

Figure P-3.2

3.3　A special and yet very common case in this family of differential equations is

$$\dot{y} + \alpha y = 0 \tag{a}$$

which is linear in y and $\dot{y} = dy/dt$, and in which α is a constant. Find the general solution (integral) of this equation and also give a graphical representation of this solution.

Solution　Such an equation is called a homogeneous linear differential equation of the first order with constant coefficients. The solution of this equation is again found by (f) of Problem 3.1 or by (d) of Problem 3.2.

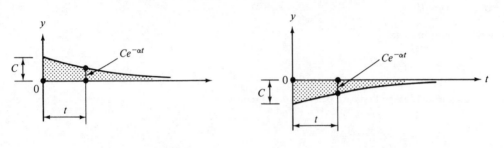

(a) Positive relaxation curve, ($C > 0$)　　　　　　(b) Negative relaxation curve, ($C < 0$)

Figure P-3.3

With $\beta = 0$, the general integral reduces to (b)

$$y = Ce^{-\alpha t} \qquad \text{(c)}$$

where α and C have the same meaning as before.

The most typical curves of this solution are the positive relaxation curve ($\alpha > 0$, $C > 0, t > 0$) and the negative relaxation curve ($\alpha > 0, C < 0, t > 0$), shown in Fig. P-3.3a and b respectively.

Initial Conditions

3.4   Compute the values $y_j = y(t_j)$ and $\dot{y}_j = \dot{y}(t_j)$ by (f) of Problem 3.1 if $y_i = y(t_i)$ is given but $\dot{y}_i = \dot{y}(t_i)$ is unknown.

Solution

The general integral (f) in Problem 3.1 can be written symbolically as

$$y = y(t) = C\psi(t) + \Phi(t) \qquad \text{(a)}$$

where

$$\psi(t) = e^{-\int P(t)\,dt} \qquad \Phi(t) = e^{-\int P(t)\,dt} \int Q(t)e^{\int P(t)\,dt}\,dt \qquad \text{(b, c)}$$

are the shape and forcing functions respectively.

At t_i, if y_i is known, then by (a) of this problem,

$$y_i = C\psi_i + \Phi_i \qquad \text{(d)}$$

where

$$\psi_i = \psi(t_i) \quad \text{and} \quad \Phi_i = \Phi(t_i).$$

Then, from (d) above,

$$C = \frac{y_i - \Phi_i}{\psi_i} \qquad \text{(e)}$$

and (a) of this problem for a given t_j, in terms of the known initial values y_i, Φ_i, ψ_i, becomes

$$y_j = \frac{\psi_i\Phi_j - \psi_j\Phi_i}{\psi_i} + \frac{y_i\psi_j}{\psi_i} \qquad \text{(f)}$$

The first time derivative of (a) at the same time station then becomes

$$\dot{y}_j = \frac{\psi_i\dot{\Phi}_j - \dot{\psi}_j\Phi_i}{\psi_i} + \frac{y_i\dot{\psi}_j}{\psi_i} \qquad \text{(g)}$$

where

$$\psi_j = \psi(t_j), \quad \dot{\psi}_j = \dot{\psi}(t_j), \quad \Phi_j = \Phi(t_j), \quad \dot{\Phi}_j = \dot{\Phi}(t_j)$$

and $t_j = t$ is variable.

3.5   Compute the values $y_j = y(t_j)$ and $\dot{y}_j = \dot{y}(t_j)$ by (f) of Problem 3.1 if $y_i = y(t_i)$ is unknown but $\dot{y} = \dot{y}(t_i)$ is given.

Solution   The first time derivative of (a) in Problem 3.4 can be written symbolically as

$$\dot{y} = \dot{y}(t) = C\dot{\psi}(t) + \dot{\Phi}(t) \tag{a}$$

where $\dot{\psi}(t), \dot{\Phi}(t)$ are the time derivatives. The analytical forms of these functions are given by (b) and (c) in Problem 3.4.

At t_i,

$$\dot{y}_i = C\dot{\psi}_i + \dot{\Phi}_i \tag{b}$$

where

$$\dot{\psi}_i = \dot{\psi}(t_i) \quad \text{and} \quad \dot{\Phi}_i = \dot{\Phi}(t_i)$$

Then from (b),

$$C = \frac{\dot{y}_i - \dot{\Phi}_i}{\dot{\psi}_i} \tag{c}$$

and (a) of the preceding problem for a given time t_j in terms of known $\dot{y}_i, \dot{\psi}_i, \dot{\Phi}_i$ becomes

$$y_j = \frac{\dot{\psi}_i \Phi_j - \psi_j \dot{\Phi}_i}{\dot{\psi}_i} + \frac{\dot{y}_i \psi_j}{\dot{\psi}_i} \tag{d}$$

In terms of the same values, (a) of this problem becomes

$$\dot{y}_j = \frac{\dot{\psi}_i \dot{\Phi}_j - \dot{\psi}_j \dot{\Phi}_i}{\dot{\psi}_i} + \frac{\dot{y}_i \dot{\psi}_j}{\dot{\psi}_i} \tag{e}$$

where all values are the time derivatives.

3.6   Reduce the results (f) and (g) of Problem 3.4 for $P(t) = \alpha$ and $Q(t) = \beta$.

Solution   If α, β are constants, then by (d) of Problem 3.2,

$$\psi(t) = e^{-\alpha t} \qquad \Phi(t) = \frac{\beta}{\alpha} \tag{a}$$

and by (e) of Problem 3.4,

$$C = \left(y_i - \frac{\beta}{\alpha}\right)e^{\alpha t} \tag{b}$$

Since $\Phi(t)$ is independent of time,

$$\Phi_i = \Phi_j = \frac{\beta}{\alpha} \qquad \dot{\Phi}_i = \dot{\Phi}_j = 0 \tag{c}$$

with which (f) of Problem 3.4 gives

$$y_j = \frac{\beta}{\alpha}(1 - e^{-\alpha \tau_{ji}}) + y_i e^{-\alpha \tau_{ji}} \qquad \text{(d)}$$

Then by (g) of Problem 3.4,

$$\dot{y}_j = \beta e^{-\alpha \tau_{ji}} - y_i \alpha e^{-\alpha \tau_{ji}} \qquad \text{(e)}$$

where $\tau_{ji} = t_j - t_i$ is the time interval ij.

3.7 Reduce the results (d) and (e) of Problem 3.5 for $P(t) = \alpha$ and $Q(t) = \beta$.

If α, β are constants, then (a) of Problem 3.6 remains   *Solution*

$$\psi(t) = e^{-\alpha t} \qquad \Phi(t) = \frac{\beta}{\alpha} \qquad \text{(a)}$$

but (c) of Problem 3.5 with $\dot{\Phi}_i = 0$ changes to

$$C = -\frac{y_i}{\alpha} e^{\alpha t_j} \qquad \text{(b)}$$

With C known, (d) of Problem 3.5 reduces to

$$y_j = \frac{\beta}{\alpha} - \frac{y_i}{d} e^{-\alpha \tau_{ji}} \qquad \text{(c)}$$

and (e) of Problem 3.5 becomes

$$\dot{y}_j = \dot{y}_i e^{-\alpha \tau_{ji}} \qquad \text{(d)}$$

where $\tau_{ji} = t_j - t_i$ is again the interval ij.

Elementary Models

3.8˙ Derive the stress-strain relationship

$$\sigma = \frac{\eta}{t}(\epsilon - \epsilon_0) \qquad \text{(a)}$$

introduced for the Newtonian model (Fig. 3-2b) as (3-2), where η is the constant of viscosity, σ is the stress, ϵ is the strain, and ϵ_0 is the strain at $t = 0$.

By the definition of the model,   *Solution*

$$\dot{\epsilon} = \frac{\sigma}{\eta} \qquad \text{(b)}$$

and after integration,

$$\epsilon = \frac{\sigma t}{\eta} + C \qquad \text{(c)}$$

The constant of integration C is computed from the initial condition $\epsilon(0) = \epsilon_0$ and is

$$C = \epsilon_0$$

With this constant known, (c) yields

$$\epsilon = \frac{\sigma t}{\eta} + \epsilon_0 \quad \text{or} \quad \sigma = \frac{\eta}{t}(\epsilon - \epsilon_0) \qquad \text{(d)}$$

which is the desired relationship (a).

For a given σ (which is a constant), the time derivative of (d) is (b).

3.9 Derive the stress-strain relationship for the Newtonian model (Fig. 3-2b), the stress function of which is a function of time.

Solution If σ is a function of t, the integration of (b) in Problem 3.8 gives

$$\epsilon(t) = \frac{1}{\eta} \int \sigma \, dt + C = \psi(t) + C \qquad \text{(a)}$$

and if at t_i,

$$\epsilon(t_i) = \epsilon_i \qquad \psi(t_i) = \psi_i$$

then

$$C = \epsilon_i - \psi_i \qquad \text{(b)}$$

where ϵ_i and ψ_i are given or computed values.

Finally, (a) in terms of (b) of this problem is

$$\epsilon(t) = \psi(t) - \psi_i + \epsilon_i \qquad \text{(c)}$$

and if at t_j,

$$\epsilon(t_j) = \epsilon_j \qquad \psi(t_j) = \psi_j,$$

then the preceding equation is

$$\epsilon_j = \psi_j - \psi_i + \epsilon_i = \psi_{ji} + \epsilon_i \qquad \text{(d)}$$

where ψ_{ji} is a specific shape function of the interval $\tau_{ji} = t_j - t_i$.

Composite Model

3.10 Derive the stress-strain relationship

$$\epsilon = \frac{\alpha_0}{k}(1 - e^{-\alpha t}) + \epsilon_0 e^{-\alpha t} \qquad \text{(a)}$$

introduced for the Kelvin model (Fig. 3-5c) as (3-5b), where η represents the constant of viscosity, k is the elastic spring constant, $\alpha = k/\eta$, ϵ is the strain at t, ϵ_0 is the strain at $t = 0$, and σ_0 is the stress at $t = 0$.

By the definition of the model, the resulting stress σ must be in equilibrium with *Solution* the stresses in the Hookean part of the model ($k\epsilon$) and in the Newtonian part of the model ($\eta\dot{\epsilon}$).

Thus

$$\eta\dot{\epsilon} + k\epsilon = \sigma \quad \text{or} \quad \dot{\epsilon} + \alpha\epsilon = \frac{\sigma}{\eta} \tag{b}$$

which is (3-5a), but also (a) of Problem 3.2, for

$$\beta = \frac{\sigma}{\eta} = \frac{\sigma_0}{\eta} \text{ constant.}$$

Then by (d) of Problem 3.6,

$$\epsilon = \frac{\sigma_0}{k}(1 - e^{-\alpha t}) + \epsilon_0 e^{-\alpha t} \tag{c}$$

which is the desired relationship (a).

3.11   Derive the stress-strain relationship

$$\sigma = e^{-\alpha t}\left(k \int \dot{\epsilon} e^{\alpha t}\, dt + \sigma_0\right) \tag{a}$$

introduced for the Maxwell model (Fig. 3-6c) as (3-6b), where η is the constant of viscosity, k is the elastic spring constant, $\alpha = k/\eta$, σ is the stress, $\dot{\epsilon}$ is the time derivative of strain, and σ_0 represents the stress at $t = 0$.

By the definition of the model, the resultant strain ϵ must be equal to the sum of *Solution* the strain in the Hookean model (ϵ_H) and the strain in the Newtonian model (ϵ_N).

The same holds true for their time derivatives; that is,

$$\dot{\epsilon} = \dot{\epsilon}_H + \dot{\epsilon}_N = \frac{\dot{\sigma}}{k} + \frac{\sigma}{\eta} \tag{b}$$

where $\dot{\epsilon}_H = \dot{\sigma}/k$ is the time derivative of (3-1) and $\dot{\epsilon}_N = \sigma/\eta$ is given by the model definition (Fig. 3-2a).

Thus

$$\dot{\sigma} + \alpha\sigma = k\dot{\epsilon} \tag{c}$$

which is (a) in Problem 3.1 with $P(t) = \alpha$ and $Q(t) = k\dot{\epsilon}$.

By (e) in Problem 3.1,

$$\sigma = e^{-\alpha t}\int k\dot{\epsilon} e^{\alpha t}\, dt + Ce^{-\alpha t} \tag{d}$$

where for $t = 0$,

$$\sigma(0) = \sigma_0 \quad \text{and} \quad C = \sigma_0$$

Then the final form of (d) becomes

$$\sigma = e^{-\alpha t}\left(k \int \dot{\epsilon} e^{\alpha t}\, dt + \sigma_0\right) \tag{e}$$

which is the desired relationship (a).

Complex Composite Models

3.12   Derive the stress-strain relationships

$$\sigma = k\epsilon \qquad (\sigma \leq \sigma_Y)$$

$$\sigma = e^{-\alpha t}\left(k \int \dot{\epsilon} e^{\alpha t}\, dt + \sigma_0\right) + \sigma_Y \qquad (\sigma \geq \sigma_Y) \tag{a}$$

introduced for the Bingham model (Fig. 3-7c) as (3-7), where η is the constant of viscosity, k is the elastic spring constant, $\alpha = k/\eta$, σ is the stress in the Hookean part of the model, σ_Y equals the friction resistance in the St. Venant part of the model, $\sigma - \sigma_Y$ represents the stress in the Newtonian part of the model, ϵ is the total strain, $\dot{\epsilon}$ is the time derivative of total strain, and σ_0 equals the stress at $t = 0$.

Solution   By the definition of the model, the resultant strain ϵ must again be equal to the sum of the strain in the Hookean model (ϵ_H) and the strain in the Newtonian model (ϵ_N).

This equality, however, is conditioned by the magnitude of σ.

If $\sigma \leq \sigma_Y$, the Newtonian part of the model cannot function and $\epsilon_N = 0$.

If $\sigma \geq \sigma_Y$, as in Problem 3.10,

$$\dot{\epsilon} = \dot{\epsilon}_H + \dot{\epsilon}_N = \frac{\dot{\sigma}}{k} + \frac{\sigma - \sigma_Y}{\eta} \tag{b}$$

where $(\sigma - \sigma_Y)/\eta$ is the strain velocity of the Newtonian part of the model with $\sigma_N = \sigma - \sigma_Y$.

Thus

$$\dot{\sigma} + \alpha\sigma = \alpha\sigma_Y + k\dot{\epsilon} \tag{c}$$

which is (a) in Problem 3.1 with $P(t) = \alpha = k/\eta$ and $Q(t) = \alpha\sigma_Y + k\dot{\epsilon}$.

By (e) in Problem 3.1,

$$\sigma = e^{-\alpha t} \int (\alpha\sigma_Y + k\dot{\epsilon})e^{\alpha t}\, dt + Ce^{-\alpha t} \tag{d}$$

where for $t = 0$,

$$\sigma(0) = \sigma_0 \qquad C = \sigma_0$$

Then the final form of (d) becomes

$$\sigma = e^{-\alpha t}\left(k \int \dot{\epsilon} e^{\alpha t}\, dt + \sigma_0\right) + \sigma_Y \qquad (e)$$

which is the desired relationship (a).

3.13   Derive the stress-strain relationship

$$\epsilon = \frac{\epsilon_0}{k_2}(1 - \beta\eta e^{-\beta k_2 t}) \qquad (a)$$

introduced for the standard linear model (Fig. 3-8c) as (3-8), where η is the constant of viscosity, k_1, k_2 are the respective spring constants, $\beta = k_1/(k_1 + k_2)$, ϵ is the total strain, and σ_0 equals the stress at $t = 0$.

By the definition of the model, the resultant stress   *Solution*

$$\sigma = \sigma_H + \sigma_M \qquad (b)$$

where σ_H = the stress in the Hookean part of the model
   σ_M = the stress in the Maxwell part of the model
   Also,

$$\epsilon = \epsilon_H = \epsilon_M \qquad (c)$$

where ϵ_H = the strain in the Hookean part of the model
   ϵ_M = the strain in the Maxwell part of the model
   By (3-1),

$$\sigma_H = k_2\epsilon \quad \text{or} \quad \epsilon = \frac{\sigma_H}{k_2} \qquad (d)$$

and by (b) in Problem 3.11,

$$\epsilon = \frac{\dot{\sigma}_M}{k_1} + \frac{\sigma_M}{\eta} \quad \text{or} \quad \sigma_M = \eta\dot{\epsilon} - \frac{\eta}{k_1}\dot{\sigma}_M \qquad (e)$$

Then (b) in terms of (d) and (e) becomes

$$\sigma = k_2\epsilon + \eta\dot{\epsilon} - \frac{\eta}{k_1}\dot{\sigma}_M \qquad (f)$$

where

$$\dot{\sigma}_M = \dot{\sigma} - \sigma_M = \dot{\sigma} - k_2\dot{\epsilon}$$

For $\sigma(t) = \sigma_0$ (constant stress), $\dot{\sigma} = 0$, and (f) reduces to

$$\dot{\epsilon} + \frac{k_1 k_2}{\eta(k_1 + k_2)}\epsilon = \frac{k_1}{\eta(k_1 + k_2)}\sigma_0 \qquad (g)$$

The general integral of (g) with

$$\beta = \frac{k_1}{\eta(k_1 + k_2)}$$

is then, according to (c) in Problem 3.2,

$$\epsilon = \underbrace{e^{-\beta k_2 t} \int \sigma_0 \beta e^{\beta k_2 t} \, dt}_{\sigma_0 / k_2} + C e^{-\beta k_2 t} \tag{h}$$

where for $t = 0$,

$$\epsilon_0 = \frac{\sigma_0}{k_1 + k_2}$$

and

$$C = -\frac{\sigma_0 k_1}{(k_1 + k_2)k_2} \tag{i}$$

With C known, the final form of (h) is

$$\epsilon = \frac{\sigma_0}{k_2}(1 - \beta \eta e^{-\beta k_2 t}) \tag{j}$$

which is the desired relationship (a).

3.14   Derive the stress-strain relationship

$$\epsilon = \sigma_0 \left[\frac{1}{k_2} + \frac{1}{\eta_2} + \frac{1}{k_1}(1 - e^{-\alpha_1 t}) \right] \tag{a}$$

introduced for the Burger model (Fig. 3-9c) as (3-9), where k_1, k_2 are the spring constants, η_1, η_2 are the dashpot constants, $\alpha_1 = k_1/\eta_1$, ϵ is the total strain, and σ_0 represents the total stress at $t = 0$.

Solution   By the definition of the model, the total strain

$$\epsilon = \epsilon_M + \epsilon_k \tag{b}$$

where $\epsilon_M =$ the resultant strain in the Maxwell part of the model
$\epsilon_k =$ the resultant strain in the Kelvin part of the model

By (b) in Problem 3.11, for $\sigma(t) = \sigma_0 =$ constant,

$$\epsilon_M = \frac{\sigma_0}{k_2} + \frac{\sigma_0 t}{\eta_2} + C_2 \tag{c}$$

where for $t = 0$,

$$\epsilon_M(0) = \frac{\sigma_0}{k_2} \quad \text{and} \quad C_2 = 0$$

By (b) in Problem 3.10, for $\sigma(t) = \sigma_0 =$ constant,

$$\epsilon_K = \frac{\sigma_0}{k_1} + C_1 e^{-\alpha_1 t} \tag{d}$$

where for $\epsilon_K(0) = 0$, $C_1 = -\sigma_0/k_1$.

The superposition of the preceding results (c) and (d) is

$$\epsilon = \epsilon_M + \epsilon_K = \sigma_0 \left[\frac{1}{k_2} + \frac{t}{\eta_2} + \frac{1}{k_1}(1 - e^{-\alpha_1 t}) \right] \tag{e}$$

which is the desired relationship (a),

Transport Matrices

3.15 Construct the time transport matrix (3-12) for the general integral (f) in Problem 3.1, relating the expanded state vector (including the time derivative)

$$\hat{Y}(t_i) = \{+1, y(t_i), \dot{y}(t_i)\} \tag{a}$$

to the expanded state vector

$$\hat{Y}(t_j) = \{+1, y(t_j), \dot{y}(t_j)\} \tag{b}$$

where $t_j > t_i$ are the coordinates on the time axis (times) of j and i respectively. Assume that y_i is known.

Solution The assembly of (f) and (g) of Problem 3.4 gives the desired time transport matrix equation; that is,

$$
\underbrace{\begin{bmatrix} +1 \\ y_j \\ \dot{y}_j \end{bmatrix}}_{\hat{Y}_j}
=
\underbrace{\begin{bmatrix} +1 & 0 & 0 \\ \dfrac{\psi_i \Phi_j - \psi_j \Phi_i}{\psi_i} & \dfrac{\psi_j}{\psi_i} & 0 \\ \dfrac{\psi_i \dot{\Phi}_j - \dot{\psi}_j \Phi_i}{\psi_i} & \dfrac{\dot{\psi}_j}{\psi_i} & 0 \end{bmatrix}}_{\hat{T}_{ji}}
\underbrace{\begin{bmatrix} +1 \\ y_i \\ \dot{y}_i \end{bmatrix}}_{\hat{Y}_i}
$$

where $\hat{Y}_i$, $\hat{Y}_j$ are the state vectors defined by (a) and (b), respectively, and $\hat{T}_{ji}$ is the time transport matrix relating the state vector $\hat{Y}_i$ to the state vector $\hat{Y}_j$.

3.16 Construct the time transport matrix for the general integral (f) in Problem 3.1 relating $\hat{Y}_i$ and, $\hat{Y}_j$, if $\dot{y}_i$ is known.

Solution The assembly of (d) and (e) of Problem 3.5 gives the desired time transport equation; that is,

$$
\underbrace{\begin{bmatrix} +1 \\ y_j \\ \dot{y}_j \end{bmatrix}}_{\hat{Y}_j}
=
\underbrace{\begin{bmatrix} +1 & 0 & 0 \\ \dfrac{\dot{\psi}_i \Phi_j - \psi_j \dot{\Phi}_i}{\dot{\psi}_i} & 0 & \dfrac{\psi_j}{\dot{\psi}_i} \\ \dfrac{\dot{\psi}_i \dot{\Phi}_j - \dot{\psi}_j \dot{\Phi}_i}{\dot{\psi}_i} & 0 & \dfrac{\dot{\psi}_j}{\dot{\psi}_i} \end{bmatrix}}_{\hat{T}_{ji}}
\underbrace{\begin{bmatrix} +1 \\ y_i \\ \dot{y}_i \end{bmatrix}}_{\hat{Y}_i}
$$

where $\hat{Y}_i, \hat{Y}_j, \hat{T}_{ji}$ have the same meaning as in Problem 3.15.

3.17 Construct the time transport matrix for the general integral (d) in Problem 3.2 relating $\hat{Y}_i$ and, $\hat{Y}_j$, if y_i or $\dot{y}_i$ is given.

Solution If y_i is known, then by (d) and (e) in Problem 3.6

$$
\underbrace{\begin{bmatrix} +1 \\ y_j \\ \dot{y}_j \end{bmatrix}}_{\hat{Y}_j} = \underbrace{\begin{bmatrix} +1 & 0 & 0 \\ \frac{\beta}{\alpha}(1 - e^{-\alpha\tau_{ji}}) & e^{-\alpha\tau_{ji}} & 0 \\ \beta e^{-\alpha\tau_{ji}} & -\alpha e^{-\alpha\tau_{ji}} & 0 \end{bmatrix}}_{\hat{T}_{ji}} \underbrace{\begin{bmatrix} +1 \\ y_i \\ \dot{y}_i \end{bmatrix}}_{\hat{Y}_i} \tag{a}
$$

If $\dot{y}_i$ is known, then by (c) and (d) of Problem 3.8

$$
\underbrace{\begin{bmatrix} +1 \\ y_j \\ \dot{y}_j \end{bmatrix}}_{\hat{Y}_j} = \underbrace{\begin{bmatrix} +1 & 0 & \\ \frac{\beta}{\alpha} & 0 & \frac{-e^{-\alpha\tau_{ji}}}{\alpha} \\ 0 & 0 & e^{-\alpha\tau_{ji}} \end{bmatrix}}_{\hat{T}_{ji}} \underbrace{\begin{bmatrix} +1 \\ y_i \\ \dot{y}_i \end{bmatrix}}_{\hat{Y}_i} \tag{b}
$$

where, in (a) and (b), $\hat{Y}_i$, $\hat{Y}_j$, $\hat{T}_{ji}$ have the same meaning as in Problem 3.15 and $\tau_{ji} = t_j - t_i$ is the time interval of $\hat{T}_{ji}$.

Transport Chains

3.18 For the Kelvin model of Fig. P-3.18a construct the transport chain for intervals $\tau_{i0} = a$, $\tau_{ji} = a$, $\tau_{kj} = a$ corresponding to the sustained load histograph of Fig. P-3.18b. Numerical values are given in Fig. P-3.18d and $\epsilon(0) = 0$.

Solution The reduced transport matrices $\hat{T}_{i0}$, $\hat{T}_{ji}$, $\hat{T}_{kj}$ for the model of Fig. P-3.18a in terms of constants of Fig. P-3.18d are, by (c) of Problem 3.10,

$$
\hat{T}_{i0} = \begin{bmatrix} +1 & 0 \\ (5)10^{-4} & 0.50 \end{bmatrix}, \quad \hat{T}_{ji} = \begin{bmatrix} +1 & 0 \\ (10)10^{-4} & 0.50 \end{bmatrix}, \quad \hat{T}_{kj} = \begin{bmatrix} +10 & 0 \\ (15)10^{-4} & 0.50 \end{bmatrix} \tag{a}
$$

The state vectors at $0, i, j, k$ are

$$
\hat{Y}_0 = \{+1, \epsilon_0\}, \quad \hat{Y}_i = \{+1, \epsilon_i\}, \quad \hat{Y}_j = \{+1, \epsilon_j\}, \quad \hat{Y}_k = \{+1, \epsilon_k\}
$$

By (3-13) and with $\epsilon(0) = \epsilon_0 = 0$,

$$
\underbrace{\begin{bmatrix} +1 \\ \epsilon_i \end{bmatrix}}_{\hat{Y}_i} = \underbrace{\begin{bmatrix} +1 & 0 \\ (5)(10)^{-4} & 0.50 \end{bmatrix}}_{\hat{T}_{i0}} \underbrace{\begin{bmatrix} +1 \\ e_0 \end{bmatrix}}_{\hat{Y}_0} = \underbrace{\begin{bmatrix} +1 \\ (5)10^{-4} \end{bmatrix}}_{\hat{Y}_i} \tag{b}
$$

$$
\underbrace{\begin{bmatrix} +1 \\ \epsilon_j \end{bmatrix}}_{\hat{Y}_j} = \underbrace{\begin{bmatrix} +1 & 0 \\ (10)10^{-4} & 0.50 \end{bmatrix}}_{\hat{T}_{ji}} \underbrace{\begin{bmatrix} +1 \\ (5)10^{-4} \end{bmatrix}}_{\hat{Y}_i} = \underbrace{\begin{bmatrix} +1 \\ (12.5)10^{-4} \end{bmatrix}}_{\hat{Y}_j} \tag{c}
$$

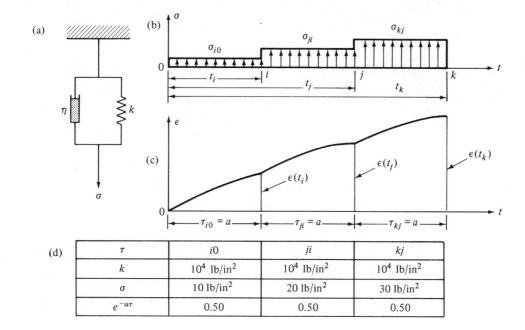

Figure P-3.18

$$\begin{bmatrix} +1 \\ \epsilon_k \end{bmatrix} = \underbrace{\begin{bmatrix} +1 & 0 \\ (15)10^{-4} & 0.50 \end{bmatrix}}_{\hat{T}_{kj}} \underbrace{\begin{bmatrix} +1 \\ (12.5)10^{-4} \end{bmatrix}}_{\hat{Y}_j} = \underbrace{\begin{bmatrix} +1 \\ (21.25)10^{-4} \end{bmatrix}}_{\hat{Y}_j} \qquad (d)$$

Obviously, if only ϵ_k is required,

$$\underbrace{\hat{Y}_k = \hat{T}_{kj}\hat{T}_{ji}\hat{T}_{i0}\hat{Y}_0}_{\hat{T}_{k0}} = \underbrace{\begin{bmatrix} +1 & 0 \\ (21.15)10^{-4} & 0.50 \end{bmatrix}}_{\hat{T}_{k0}} \underbrace{\begin{bmatrix} +1 \\ \epsilon_0 \end{bmatrix}}_{\hat{Y}_0} = \underbrace{\begin{bmatrix} +1 \\ (21.25)10^{-4} \end{bmatrix}}_{\hat{Y}_k}$$

where $\hat{T}_{k0}$ is a product of $\hat{T}$-matrices of (d), (c), and (b), respectively.

Although the same results can be obtained by solving the differential equation for each interval, the matrix solution is simple, mechanical, and adjustable to any number of intervals.

It must be noted that for equal-span intervals, ψ functions are the same but Φ changes with the loading.

3.19   Consider the Kelvin model of Problem 3.18 for $\sigma_{i0} = 10\,\mathrm{lb/in^2}$, $\sigma_{ji} = \sigma_{kj} = 0$. Assume the same constants and also $\epsilon_0 = 0$.

The transport matrices (a) of the preceding problem change their load terms in   *Solution* τ_{ji} and τ_{kj} to zero; that is,

$$\hat{T}_{i0} = \begin{bmatrix} +1 & 0 \\ (5)10^{-4} & 0.50 \end{bmatrix}, \quad \hat{T}_{ji} = \begin{bmatrix} +1 & 0 \\ 0 & 0.50 \end{bmatrix}, \quad \hat{T}_{kj} = \begin{bmatrix} +1 & 0 \\ 0 & 0.50 \end{bmatrix} \tag{a}$$

and $\hat{Y}_i = \{+1, (5)10^{-4}\}$ remains the same.

But

$$\hat{Y}_j = \underbrace{\begin{bmatrix} +1 & 0 \\ 0 & 0.50 \end{bmatrix}}_{\hat{T}_{ji}} \underbrace{\begin{bmatrix} +1 \\ (5)10^{-4} \end{bmatrix}}_{\hat{Y}_i} = \underbrace{\begin{bmatrix} +1 \\ (2.5)10^{-4} \end{bmatrix}}_{\hat{Y}_j}$$

and

$$\hat{Y}_k = \underbrace{\begin{bmatrix} +1 & 0 \\ 0 & 0.50 \end{bmatrix}}_{\hat{T}_{kj}} \underbrace{\begin{bmatrix} +1 \\ (0.25)10^{-4} \end{bmatrix}}_{\hat{Y}_j} = \underbrace{\begin{bmatrix} +1 \\ (0.125)10^{-4} \end{bmatrix}}_{\hat{Y}_k}$$

Thus, as the load is removed, the model's strain recedes along the relaxation curve.

SUPPLEMENTARY PROBLEMS

General Integrals ($\alpha, \beta, A, B, C = $ *Constants*)

3.20   Solve $\dot{y} + 2ty = 4t$, using the integrating factor.

Answer $y = 2 + Ce^{-t^2}$

3.21   Solve $\dot{y} + By = Q(t)$, using a series.

Answer $y = Ce^{-Bt} + \dfrac{Q(t)}{B} - \dfrac{\dot{Q}(t)}{B^2} + \dfrac{\ddot{Q}(t)}{B_3} - \cdots$

3.22   Solve $\dot{y} + By = Ae^{\alpha t}$, using the integrating factor.

Answer $y = Ce^{-Bt} + \dfrac{A}{\alpha + B} e^{\alpha t}$

3.23   Solve $\dot{y} + By = Ae^{\alpha t} \sin \beta t$, using the integrating factor.

Answer $y = Ce^{-Bt} - \dfrac{A[(\alpha + B) \sin \beta t + \beta \cos \beta t]}{\beta^2 + (\alpha + B)^2}$

3.24   Solve $\dot{y} + By = Ae^{\alpha t} \cos \beta t$, using the integrating factor.

Answer $y = Ce^{-\beta t} + \dfrac{A[\beta \sin \beta t + (\alpha + B) \cos \beta t]}{\beta^2 + (\alpha + B)^2}$

Initial Conditions

3.25   Compute C in Problem 3.20, given $y(0) = 0$.

$C = -2$ *Answer*

3.26 Compute C in Problem 3.20, given $y(t_i) = y_i$.

$C = (y_i - 2)e^{t_i^2}$ *Answer*

3.27 Compute C in Problem 3.21, given $y(0) = 0, Q(0) = B, \dot{Q}(0) = \ddot{Q}(0) = \cdots = 0$.

$C = -1$ *Answer*

3.28 Compute C in Problem 3.22, given $y(0) = (A + B)/(\alpha + B)$.

$C = \dfrac{B}{\alpha + B}$ *Answer*

3.29 Compute C in Problem 3.24, given $y(0) = 0$.

$C = \dfrac{-A(\alpha + B)}{\beta^2 + (\alpha + B)^2}$ *Answer*

Transport Matrices

3.30 Construct the time transport matrix for the Newtonian model (Fig. 3-2c), if $\epsilon(t_i) = \epsilon_i$ and $\sigma(t_i) = \sigma_i$ are known.

$$\hat{T}_{ji} = \begin{bmatrix} +1 & 0 \\ \dfrac{\sigma_i t_i}{\eta} & +1 \end{bmatrix}, \qquad \begin{matrix} \hat{Y}_i = \{+1, \epsilon_i\} \\ \hat{Y}_j = \{+1, \epsilon_j\} \end{matrix}$$ *Answer*

3.31 Construct the time transport matrix for the Maxwell model (Fig. 3.6c), if $\dot{\epsilon} = a = $ constant and $\sigma(0) = \sigma_0 = $ constant are known.

$$\hat{T}_{ji} = \begin{bmatrix} +1 & 0 \\ \dfrac{ak}{\alpha}(1 - e^{-\alpha \tau_{ji}}) & e^{-\alpha \tau_{ji}} \end{bmatrix}, \qquad \begin{matrix} \hat{Y}_i = \{+1, \sigma_i\} \\ \hat{Y}_j = \{+1, \sigma_j\} \end{matrix}$$ *Answer*

3.32 Construct the time transport matrix for the Burger model (Fig. 3-9c), if $\sigma(t) = \sigma_0 = $ constant and $\epsilon(t_i) = \epsilon_i$ are known.

$$\hat{T}_{ji} = \begin{bmatrix} +1 & 0 & 0 \\ \dfrac{\alpha_0 \tau_{ji}}{\eta_2} & +1 & 0 \\ \dfrac{\alpha_0}{k_1}(1 - e^{-\alpha_1 \tau_{ji}}) & 0 & e^{-\alpha_1 \tau_{ji}} \end{bmatrix}, \qquad \begin{matrix} \hat{Y}_i = \{+1, \epsilon_{Mi}, \epsilon_{Ki}\} \\ \\ \hat{Y}_j = \{+1, \epsilon_{Mj}, \epsilon_{Kj}\} \end{matrix}$$ *Answer*

Note: Since the Burger model is nonlinear, ϵ_M and ϵ_K are kept as two independent vector components and summed for each time station.)

Transport Chains

3.33 Using the time transport chain, investigate the behavior of the Kelvin model of Fig. 3-5c for the load histograph of Fig. P-3.33a.

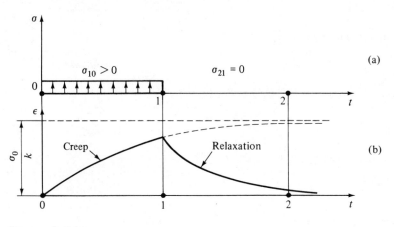

(b)

Figure P-3.33

Answer Figure P-3.33b

3.34 Using the time transport chain, investigate the behavior of the standard linear model of Fig. 3-8c for the load histograph of Fig. P-3.34a.

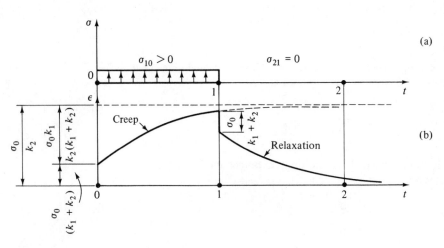

(a)

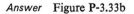

(b)

Figure P-3.34

Answer Figure P-3.34b

3.35 Using the time-transport chain, investigate the behavior of the Burger model of Fig. 3-9c for the load histograph of Fig. P-3.35a.

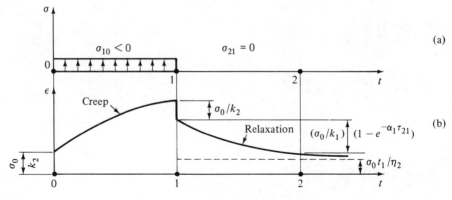

(a)

(b)

Figure P-3.35

Figure P-3.35b

Answer

soil
water
at
rest
4

4.1 INTRODUCTION

General

The *moisture in soil*, next to the texture of soil, is its most important characteristic. Soil properties such as consistency, strength, and stability are functions of the water content and change with its variation. Furthermore, the properties of the soil-water system greatly influence the value of the coefficient of viscosity η and must be known in the construction of rheological models (Chapter 3).

Classification

Two forms of moisture can occur in soil:

1. The *adsorbed water* surrounding the grains in the form of a viscous film.
2. The *free water* filling partially or completely the voids between the grains.

The free water is *at rest* (stationary, in the state of static equilibrium) or *in motion* (flow, in the state of dynamic equilibrium). Only the stationary free water is considered in this chapter.

Models

If the voids are completely filled with water (saturated soil), the moisture is said to be *continuous*, and the model representation is rather simple. If, however, the voids are only partially filled, the moisture is *discontinuous*, forming wedges between grains and water films around the grains. The model representation of this state is more involved and is the prime concern of the subsequent discussion.

Methods

Two concepts of analysis of soil water have been developed:

1. The *static equilibrium method*, also known as the capillary rise method, based on the principles of static equilibrium in capillary tubes (Section 4.2).
2. The *energy method*, also known as the fluid potential method, based on the energy potential of the soil water (Section 4.3).

Constants

Unless otherwise specified, the soil moisture considered in this chapter is the *distilled water of unit weight* $\gamma_w = 62.43 \text{ lb/ft}^3$ or 1 g/cm^3 at $4°C$. The values of γ_w at various temperatures are given in Table 4–1. Other constants (pertinent to the study of soil moisture) and their conversion factors are recorded in Table 4–2.

Unit Weight of Water in Grams Per Cubic Centimeter TABLE 4-1

°C	0	1	2	3	4	5	6	7	8	9
0	0.9999	0.9999	1.0000	1.0000	1.0000	1.0000	1.0000	0.9999	0.9999	0.9998
10	0.9997	0.9996	0.9995	0.9994	0.9993	0.9991	0.9990	0.9988	0.9986	0.9984
20	0.9982	0.9980	0.9978	0.9976	0.9973	0.9971	0.9968	0.9965	0.9963	0.9960
30	0.9957	0.9954	0.9951	0.9947	0.9944	0.9941	0.9937	0.9934	0.9930	0.9926
40	0.9922	0.9919	0.9915	0.9911	0.9907	0.9902	0.9898	0.9894	0.9890	0.9885
50	0.9881	0.9876	0.9872	0.9867	0.9862	0.9857	0.9852	0.9848	0.9842	0.9838
60	0.9832	0.9827	0.9822	0.9817	0.9811	0.9806	0.9800	0.9795	0.9789	0.9784
70	0.9778	0.9772	0.9767	0.9761	0.9755	0.9749	0.9743	0.9737	0.9731	0.9724
80	0.9718	0.9712	0.9706	0.9699	0.9693	0.9686	0.9680	0.9673	0.9667	0.9660
90	0.9653	0.9647	0.9640	0.9633	0.9626	0.9619	0.9612	0.9605	0.9598	0.9591

International Critical Tables, *Vol. 3, McGraw-Hill, New York, 1928.*

TABLE 4-2   *Conversion Factors*

---

1. Temperature
    1° Centigrade (C) = 1.8° Fahrenheit (F) = 0.8° Reaumur (Re)
    $\qquad\qquad$ = 1.8° Rankine (Ra) = 1° Kelvin (K)
    0°C = 32°F = 0°Re = 492°Ra = 273°K
    100°C = 212°F = 80°Re = 672°Ra = 373°K

---

2. Density
    1 lb/ft³ = 5.787 × 10⁻⁴ lb/in² = 16.018 × kg/cu meter = 16.018 × 10⁻² gram/cm³
    1 gram/cm³ = 0.03613 lb/in² = 62.43 lb/ft³

---

3. Pressure
    1 atmosphere (atm) = 1.0133 bars = 14.696 lb/in³. = 1.013246 × 10⁻⁶ dynes/sq cm
    $\qquad$ = 1,033.2 grams/cm² (0°C) = 760 mm Hg (0°C) = 29.92 in. Hg (0°C) = 33.903 ft water (0°C)
    1 dyne/cm² = 1.01971 × 10⁻³ gram/cm² = 1.4504 × 10⁻⁵ lb/in.
    1 bar = 1.0 × 10⁶ dynes/cm² = 0.98692 atm
    1 lb wt/in². = 70.307 grams/cm² = 68.947 dynes/cm²

---

4. Force
    1 newton = 1 × 10⁵ dynes = 0.22481 lb wt
    1 dyne = 2.2481 × 10⁻⁶ lb wt = 7.2330 × 10⁻⁵ poundal = 0.0010197 gram wt
    1 gram wt = 0.07932 poundal = 980.665 dynes = 2.2046 × 10⁻³ lb wt
    1 lb wt = 32.174 poundals = 453.59 gram wt = 1.4482 newtons
    1 pundal = 0.031081 lb wt = 14.098 gram wt = 1.3825 × 10⁴ dynes

---

Adapted in part, from William D. Cockrell, Industrial Electronics Handbook, *McGraw-Hill, New York, 1958. For more extensive tables of conversion factors refer to E. A. Mechtly, Physical Constants and Conversion Factors,* Scientific and Technical Division, NASA, *Publication SP-7012, Washington, D.C., 1964.*

4.2 STATIC EQUILIBRIUM METHOD

Capillary Rise in Glass Tube

The moisture retained between soil particles is held in position by the *surface tension* around the contact points or in the soil pores and capillaries (Fig. 4-1). The oldest approach to the analysis of this phenomenon is based on the model repesenting the soil pores as a *system of capillary tubes.* Their capillary rise h (Fig. 4-2) is then given as

$$h = \frac{2T_s \cos \alpha}{r\gamma_w} \qquad (4\text{-}1)$$

where $\quad T_s$ = surface tension
$\qquad\quad \alpha$ = angle of contact
$\qquad\quad r$ = radius of tube
$\qquad\quad \gamma_w$ = unit weight of water

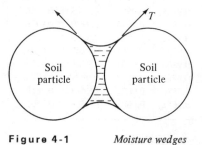

Figure 4-1   *Moisture wedges between particles*

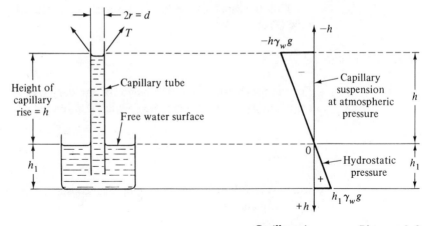

Capillary rise   **Figure 4-2**

The derivation of this equation, based on the static equilibrium of the weight of the suspended water column in the tube ($\pi r^2 h \gamma_w$) and the vertical component of the surface tension resultant ($2\pi r T \cos \alpha$), is shown in Problem 4.1.

For glass and water, $\alpha = 0$, $T \cong 0.074$ g/cm (Table 4-3, at 20°C), $\gamma_m = 1$ g/cm³, and the maximum capillary rise is

$$h_{cmax} = \frac{0.296}{d} \cong \frac{0.3}{d} \qquad (4\text{-}2)$$

where d represents the diameter of the tube.

Variation of Surface Tension of   **TABLE 4-3**
Water with Temperature

Temperature C°	Surface Tension (g/cm)
−5	0.07791
0	0.77713
5	0.07640
10	0.07567
15	0.07494
20	0.07418
25	0.07339
30	0.07258
35	0.07177
40	0.07093
100	0.06001

Thus the capillary rise h_{cmax} is inversely proportional to the diameter of the tube (Problem 4.2).

Hazen's Formula

For the rough estimate of the height of capillary rise, A. Hazen developed an empirical formula (Problem 4.3), which is given below.

$$h = \frac{C}{eD_{10}} \tag{4-3}$$

where $C = 0.1$ to 0.5 cm^2 is an empirical constant, e is the void ratio, and D_{10} is already known (2-2).

Capillary Rise in Soil

The capillary tube model of the preceding articles is a useful visualization of the soil-moisture action, yet it is an unrealistic model representation. In soils, there are no regular capillary tubes as such. Rather, the channels connecting the voids in the soils are irregular in shape, size, and distribution. These conditions limit the use of the static equilibrium method only to the calculation of water content above the ground water table.

4.3 ENERGY METHOD

Potential Energy

Since the static equilibrium method proved to be of limited value, it was only logical to choose the second possibility and approach the soil-water problem from the potential energy concept.

The introduction of the energy method in soil mechanics dates back to 1907 when E. Buckingham formulated the capillary potential as the work per unit volume of water required to pull water away from soil.

Soil-Suction Method

The concept of capillary potential led eventually to the development of the soil-suction method. In its application, the soil-suction is first determined experimentally and expressed in pF-values, where pF is the common logarithm of a certain length of a suspended water column in centimeters of water. Then the relationship between pF and the equivalent suction (negative hydraulic head) is obtained from a chart or a conversion table (Table 4-4).

Soil-Suction Coefficients   **TABLE 4-4**

pF	Equivalent Negative Hydraulic Head (cm)	Suction (lb/in².)
0	1	$(1.42)10^{-2}$
1	10	$(1.42)10^{-1}$
2	10^2	$(1.42)10$
3	10^3	$(1.42)10^1$
4	10^4	$(1.42)10^2$
5	10^5	$(1.42)10^3$
6	10^6	$(1.42)10^4$

Laboratory Measurements

Several methods are available for the measurements of soil suction, depending on the range of the required suction magnitude. For example, the *suction plate method*, developed by the Road Research Laboratory in England (Problem 4.5) is being used for suction measurements at less than 12 lb/in.² or about 3 pF (Problems 4.4 through 4.9).

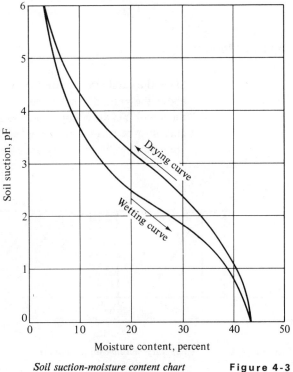

Soil suction-moisture content chart   **Figure 4-3**

The graphical representation of the soil suction-moisture content relationship is given in Fig. 4-3. After this relationship has been determined in the laboratory for a particular soil, the moisture content at any desirable level above the water table can be determined by reading off the corresponding value of moisture at the suction value equivalent to this height (Problem 4.10).

4.4 STATIC EFFECTS OF SOIL MOISTURE

Intergranular and Pore-Water Pressure

According to the principle of action and reaction, the pressure must be *transmitted throughout the medium*. The agents of transmission are the *soil grains* and the *porewater*, which in turn are subjected to forces designated as the intergranular pressures and the pore-water pressures respectively.
Their definitions are

1. The *intergranular pressure*, sometimes called the effective pressure, is defined as the contact stress between the mineral grains of soil.
2. The *pore-water pressure*, frequently designated as the hydrostatic or neutral pressure, is the hydrostatic stress in the pores.

Capillary Moisture Load

Since the capillary moisture is held suspended in a state of negative pressure above the ground water table, it does not affect the pore-water pressure below the table. Yet at the same time, as a gravity load, it must rest on the soil below and must be treated as an applied load, the magnitude and variation of which are given by the load diagram of Fig. 4-4 (Problem 4.12).

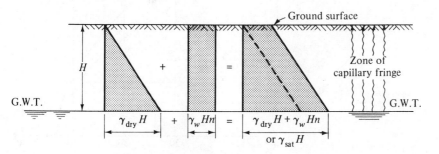

Figure 4-4   *Capillary moisture load diagram*

Superposition of Effects

The superposition of the intergranular pressure, pore-water (hydrostatic) pressure, and the capillary moisture load (if present) yields the *total pressure* in the given soil system at the selected level. Tables 4-5 and 4-6 illustrate the algebraic formulation of this problem for two typical soil profiles (Problems 4.11 to 4.14).

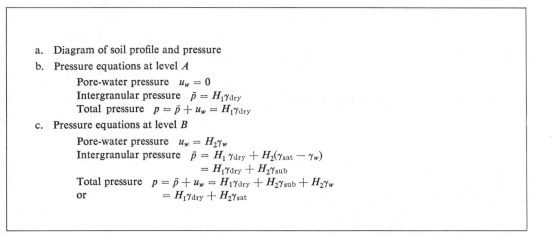

Intergranular and Pore-Water Pressure   **TABLE 4-5**
Dry Sand above Ground Water Table

a. Diagram of soil profile and pressure
b. Pressure equations at level A

    Pore-water pressure $u_w = 0$
    Intergranular pressure $\bar{p} = H_1 \gamma_{dry}$
    Total pressure $p = \bar{p} + u_w = H_1 \gamma_{dry}$

c. Pressure equations at level B

    Pore-water pressure $u_w = H_2 \gamma_w$
    Intergranular pressure $\bar{p} = H_1 \gamma_{dry} + H_2(\gamma_{sat} - \gamma_w)$
                         $= H_1 \gamma_{dry} + H_2 \gamma_{sub}$
    Total pressure $p = \bar{p} + u_w = H_1 \gamma_{dry} + H_2 \gamma_{sub} + H_2 \gamma_w$
    or $= H_1 \gamma_{dry} + H_2 \gamma_{sat}$

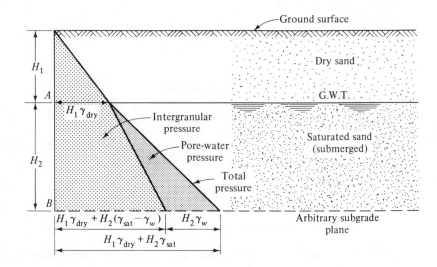

TABLE 4-6   *Intergranular and Pore-Water Pressure*
Capillary Fringe above Ground Water
Table

a. Diagram of soil profile and pressure
b. Pressure equations at level A

   Pore-water pressure $u_w = 0$
   Intergranular pressure $\bar{p} = (H_1\gamma_{\mathrm{dry}} + H_1\gamma_w\eta)$
   (where η = porosity) or $\bar{p} = H_1\gamma_{\mathrm{sat}}$
   Total pressure $p = \bar{p} + u_w = H_1\gamma_{\mathrm{sat}}$

c. Pressure equations at level B

   Pore-water pressure $u_w = H_2\gamma_w$
   Intergranular pressure $\bar{p} = H_1\gamma_{\mathrm{sat}} + H_2(\gamma_{\mathrm{sat}} - \gamma_w)$
   $\qquad\qquad\qquad\qquad = H_1\gamma_{\mathrm{sat}} + H_2\gamma_{\mathrm{sub}}$
   Total pressure $p = \bar{p} + u_w = H_1\gamma_{\mathrm{sat}} + H_2\gamma_{\mathrm{sub}} + H_2\gamma_w$
   or $\qquad\qquad\qquad = (H_1 + H_2)\gamma_{\mathrm{sat}}$

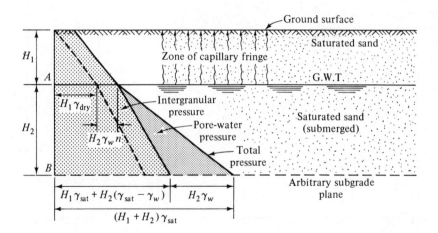

SOLVED PROBLEMS

Static Equilibrium Method

4.1 Derive the static equilibrium equation (4-1) for the height of rise h of liquid in a capillary tube of radius r shown in Fig. 4-2.

Solution The weight of the suspended water column inside the tube of Fig. 4-2, acting downward, is

$$W_w = \pi r^2 h\gamma_w \ \downarrow$$

and the resultant of the surface tension force, acting upward, is

$$F_s = T_s 2\pi r \cos\alpha \ \uparrow$$

These two forces must be in a state of static equilibrium ; that is,

$$-\pi r^2 h \gamma_w + T_s 2\pi r \cos \alpha = 0$$

From this condition,

$$h = 2T_s \cos \frac{\alpha}{\gamma_w r}$$

4.2   Three capillary tubes with diameters $d_1 = d$, $d_2 = 1.5d$, $d_3 = 2d$ were dipped in a free-water surface. Compute the heights of their capillary rise h_1, h_2, h_3.

By (4-1), the respective heights are   *Solution*

$$h_1 = 4T_s \frac{\cos \alpha}{d_1 \gamma_w}, \quad h_2 = 4T_s \frac{\cos \alpha}{d_2 \gamma_w}, \quad h_3 = 4T_s \frac{\cos \alpha}{d_3 \gamma_w}$$

and their proportions

$$h_1 : h_2 : h_3 = 1 : \frac{2}{3} : \frac{1}{2}$$

yield

$$h_1 = h, \quad h_2 = \frac{2h}{3}, \quad h_3 = \frac{h}{2}$$

which demonstrates that the height of the capillary rise is inversely proportional to the diameter of the respective tube, as shown in Fig. P-4.2.

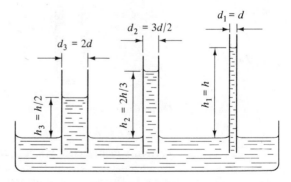

Figure P-4.2

4.3   The effective grain size of a uniform fine sand is $D_{10} = 0.052$ mm and its void ratio is $e = 0.65$. Using Hazen's formula (4-3), calculate the height h_c of capillary rise in this sand for $C = 0.1$ cm^2 and $C = 0.5$ cm^2.

Solution By (4-3), for $C = 0.1$ cm^2.

$$h_c = \frac{0.1}{(0.63)(0.0052)} = 30.5 \text{ cm}$$

and for $C = 0.5$ cm^2,

$$h_c = \frac{0.5}{(0.63)(0.0052)} = 152.5 \text{ cm}$$

The upper limit of C is applicable when sand grains are rough and not very pure, whereas the lower limit is used in sand with clean and rounded particles.

Energy Method

4.4 Describe the tensiometer method for measuring soil-suction pressure.

Answer The tensiometer apparatus is shown in Fig. P-4.4. It consists of a porous pot placed in the soil and filled with water. The pot is then connected by a water-filled tube to

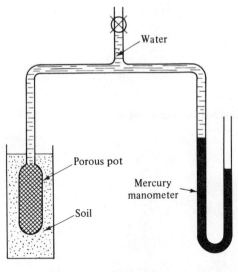

Figure P-4.4   *The tensiometer apparatus*

a mercury manometer. Water is drawn through the pores of the suction pressure of the surrounding soil until an equilibrium pressure inside and outside the pot is reached. The equilibrium suction pressure inside the pot is deduced from the mano-meter reading, while the moisture content of the soil surrounding the pot is measured in the usual manner. The soil-suction—moisture-content relationship is explored by using a range of initial soil moisture contents, and a wetting curve similar to that in Fig. 4-3 is drawn.

4.5 Describe the direct suction method for measuring soil-suction pressure.

The apparatus used in the direct suction method is shown in Fig. P-4.5. In this   *Answer*
method, a thin sample of the saturated soil is placed in a Büchner-type funnel and

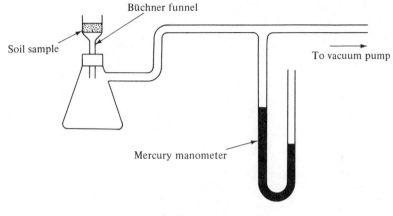

The direct suction apparatus   Figure P-4.5

subjected to a certain magnitude of suction pressure. At this pressure, the water
leaves the soil until the soil suction rises to a value equal to the applied suction.
When a stable condition is reached, one at which no additional water is drawn away
from the soil, the moisture content of the soil is determined. By using a range of
applied suction pressure, the soil-suction—moisture-content relationship in a drying
process, similar to that in Fig. 4-3, can be deter-
mined.

4.6 Describe the suction plate method for measuring soil-suction pressure.

The apparatus used in the suction plate method is   *Answer*
illustrated in Fig. P-4.6. The test procedure is
very similar to that used in the direct suction
method (Problem 4.5), except that the suction
pressure is provided in the plate hydrostatically.
The soil sample is placed on the plate, and a transfer of moisture takes place until the plate and the
sample are in a suction equilibrium. The vacuum
pump and its associated manometer (Fig. P-4.6)
are used to control the suction in the plate and
hence the vapor pressure in the enclosure.

4.7 Describe the centrifuge method for measuring soil-suction pressure.

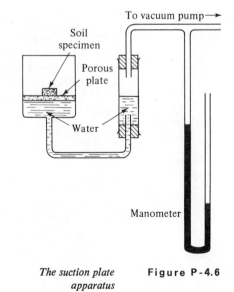

The apparatus used in the centrifuge method is   *Answer*

The suction plate   Figure P-4.6
apparatus

illustrated in Fig. P-4.7. In this method, a thin section of the soil sample is placed in a cup of the type shown in the figure. After applying a centrifuging force, water

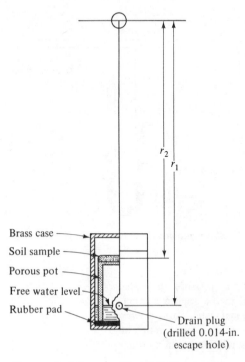

Brass case
Soil sample
Porous pot
Free water level
Rubber pad
Drain plug
(drilled 0.014-in.
escape hole)

Figure P-4.7   *The centrifuge apparatus*

is separated from the soil and travels through the porous walls of the pot to join the water table, which is maintained at a constant level during the test. When equilibrium pressure is reached, the soil-suction pressure is sufficient to prevent any further moisture from leaving the sample.

The soil-suction pressure h_p is then calculated from the speed of the centrifuge, using the formula

$$h_p = \frac{\omega^2}{2g}(r_1^2 - r_2^2)$$

where h_p = suction pressure expressed in terms of the height of the equivalent water column expressed in centimeters (i.e., log h = pF of soil moisture) (Section 4.3, page 79.)

ω = angular velocity

r_1, r_2 = the distances of the water table and the center of the soil sample from the center of rotation respectively (both measured in centimeters)

By carrying out tests at several centrifuge speeds, the relationship between soil suction and moisture content can be determined.

4.8 Compare the four methods used in determining soil-suction pressure and described in Figs. P-4.4 to P-4.7. Also, indicate some of the limitations and advantages of each method.

Comparisons, limitations, and advantages of these four methods are   *Answer*

1. The tensiometer method (Problem 4.4) is suitable only as a wetting method and for suctions up to about 12 lb/in^2.
2. The direct suction method (Problem 4.4) is suitable only as a drying method. Similar to the tensiometer method, it is only suitable for measuring suctions up to about 12 lb/in^2. Unless the soil specimen is covered during the test, evaporation from the surface may take place thereby leading to erroneous results. Also, the soil sample is subjected to the atmospheric pressure acting only on the upper face of the specimen.
3. The suction plate method (Problem 4.6) is also suitable for measuring suctions up to about 12 lb/in^2., for the porosity of the plate (Fig. P-4.6) is such that the air will not pass through when it is being used for suctions less than about 12 lb/in^2. The method is applicable to wetting and drying tests.
4. The centrifuge method (Problem 4.7) has the advantage over all other methods in that suction pressures of several hundred pounds per square inch can be obtained. However, during the centrifuging process, the soil sample is subjected to an overburden pressure because of its weight; at high speeds this pressure may become considerable. It is for this reason that very thin samples should be used. This test is only suitable for a drying process.

4.9 Enumerate the factors governing the soil-suction—moisture-content relationship and explain the influence of each factor.

From typical results of soil suction—moisture content relationships determined by   *Answer*
any of the methods in Problems 4.4 to 4.7, the following factors were found to have a significant influence on the results:

1. The clay content. The soil moisture retained in the soil under a certain suction pressure increases with the clay content of the soil. The reason is that the number of smaller channels in which water is retained as well as the total surface area of the particles both increase with the increase of the percentage of finer particles of soil.
2. The soil structure and density. Although the effect of these factors appears to have an influence on the soil suction—moisture content relationship, their role has not been fully investigated. Therefore tests to determine soil-suction pressure should be performed on undisturbed samples.
3. The soil temperature. With increasing temperature, the surface tension at the air-water interface decreases slightly, which would cause a reduction in soil-suction pressure. However, the effect of this factor is small and, for all practical purposes, can be neglected.

4.10 Determine the equilibrium moisture distribution in a mass of sandy-clay soil having a suction-moisture content relationship as shown in Fig. P-4.10a. The soil

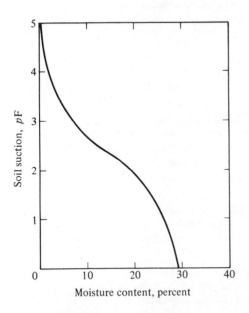

Moisture content, percent

Figure P-4.10a

porosity is $n = 0.38$, the submerged unit weight is $\gamma_{sub} = 61$ lb/ft^3, the specific gravity is $G_s = 2.70$, and the zone of soil above the ground water table is assumed to be saturated with capillary moisture. Since the density is expected to vary with depth (depending on the increase in effective pressure), in order to simplify this problem, an initial average constant value of density $\gamma_{dry} = 100$ lb/ft^3 is assumed.

Solution First, certain values must be calculated as follows:

1. The surface tension produced by the capillary rise of water in the top 6 feet is T_s (see Fig. 4-4), given as

$$T_s = \gamma_w Hn = (62.4)(6)(0.38) = 142 \text{ lb/ft}^2$$

where H is the depth of capillary saturated zone of soil.

2. The effective pressure at any point above G.W.T. is

$$\bar{p}_1 = \gamma_{dry} z_1 + T_s = 100(z_1 + 1.42) \tag{a}$$

where z_1 is the vertical coordinate of a point below the ground surface within the saturated capillary zone.

3. The effective pressure at any point below G.W.T. is

$$\bar{p}_2 = \gamma_{dry} H + T_s + \gamma_{sub} z_2 = 100(6 + 1.42) + 61 z_2 \tag{b}$$

where z_2 is the vertical position coordinate of a point below G.W.T.

Now, with these values known, the procedure for the computation of the first approximation for the equilibrium moisture content with depth consists of the following steps:

1. First the values of $\bar{p}_1$ and $\bar{p}_2$ are computed by equations (a) and (b), respectively, and recorded in the $\bar{p}$ column of Table P-4.10a.

2. Next, the suction pressure is computed as $S = 0.49\bar{p}$ and recorded in the column S.

3. Then the value of pF as log S is listed in the column pF.

4. Once pF is known for each depth z, the corresponding water content w in percent is read from Fig. P-4.10a and recorded in the w column.

5. Finally, the dry and submerged densities of soil (above and below G.W.T. respectively) are calculated by

$$\gamma_{\text{dry}} = \frac{G_s}{1 + wG_s}\gamma_w \quad \text{and} \quad \gamma_{\text{sub}} = \gamma_{\text{sat}} - \gamma_w = \frac{G_s - 1}{1 + wG_s}\gamma_w$$

TABLE P-4.10a   **TABLE P-4.10b**

FIRST APPROXIMATION						SECOND APPROXIMATION				
z	$\bar{p}$	S	pF	w	γ	z	$\bar{p}$	S	pF	w
(ft)	(lb/ft²)	(g/cm²)	log S	%	(lb/ft³)	(ft)	(lb/ft²)	(g/cm²)	log S	%
0	142	70	1.85	18	113	0	142	70	1.85	21.0
2	342	168	2.23	16	118	2	368	180	2.26	17.0
4	542	266	2.43	13	125	4	604	296	2.47	14.0
6	742	364	2.56	12	127	6	854	419	2.62	10.5
8	864	423	2.63	11	82	8	1110	544	2.74	9.5
10	986	483	2.68	10	84	10	1274	624	2.80	9.0
12	1108	543	2.74	9	85	12	1442	707	2.85	8.5
20	1596	782	2.89	8	87	20	2133	1047	3.02	7.0

The procedure of computing the second approximations (Table P-4.10b) consists of a repetition of the preceding steps; but in terms of the new unit weights listed in the γ column of the first approximations (Table P-4.10a). This process may be repeated until the γ values stabilize.

The equilibrium moisture distribution versus the depth are then plotted from the last approximation, as shown in Fig. P-4.10b. Two approximations usually yield a good result.

Static Effects of Soil Moisture

4.11 Compute the vertical effective (intergranular) and total pressures at levels A, B, and C in the mass of soil represented by Fig. P-4.11a. The sandy-loam material

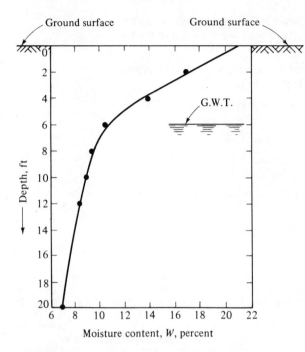

Figure P-4.10b

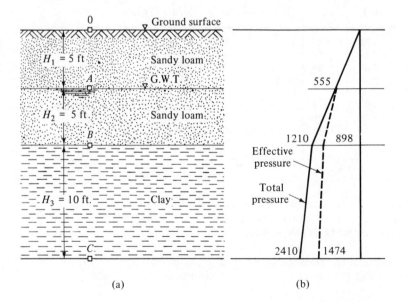

(a)   (b)

Figure P-4.11

in the top 10 feet has a dry density $\gamma_{dry} = 111$ lb/ft³ and a saturated density $\gamma_{sat} = 131$ lb/ft³. The clay material at the bottom has a saturated density $\gamma_{sat} = 120$ lb/ft³.

Assume no capillary moisture above G.W.T.

By Table 4-5, the pressures are computed at the three levels as shown below.   *Solution*

Pressure at level A:

Pore-water pressure	0	0
Intergranular pressure	$H_1 \gamma_{dry}$	555 lb/ft²
Total pressure		555 lb/ft²

Pressure at level B:

Pore-water pressure	$H_2 \gamma_w$	312 lb/ft²
Intergranular pressure	$H_1 \gamma_{dry} + H_2 \gamma_{sub}$	898 lb/ft²
Total pressure		1210 lb/ft²

Pressure at level C:

Pore-water pressure	$(H_2 + H_3)\gamma_w$	936 lb/ft²
Intergranular pressure	$H_1 \gamma_{dry} + H_2 \gamma_{sub} + H_3 \gamma_{sub}$	1474 lb/ft²
Total pressure		2410 lb/ft²

The pressure diagram for this soil profile is shown in Fig. P-4.11b.

4.12 Solve Problem 4.11 assuming that the top 5 feet of soil above G.W.T. is saturated with capillary moisture.

In this case, the capillary moisture will cause an increase in the effective pressure   *Solution*
at all levels (see Section 4.4 and Fig. 4.4). This increase in the effective pressure is
Δp, and

$$\Delta p = 5(\gamma_{sat} - \gamma_{dry}) = 5(131 - 111) = 100 \text{ lb/ft}^2$$

The effective pressures at A, B, C, denoted as $\bar{p}_A$, $\bar{p}_B$, $\bar{p}_C$, respectively, are

$$\bar{p}_A = 555 + 100 = 655 \text{ lb/ft}^2$$
$$\bar{p}_B = 898 + 100 = 998 \text{ lb/ft}^2$$
$$\bar{p}_C = 1474 + 100 = 1574 \text{ lb/ft}^2$$

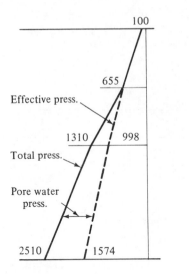

Figure P-4.12

The pore-water (hydrostatic) pressure will remain the same as in Problem 4.11, and the total pressure is then the sum of the pore-water pressure and the effective pressure. The pressure diagram for this case is shown in Fig. P-4.12.

4.13 Exploration at a building site revealed the soil profile shown in Fig. P-4.13. The water table level was originally at the ground surface. It was then lowered by drainage to a level 20 ft below ground surface. Calculate the vertical effective pressure at point C before and after lowering the G.W.T. The sand above the water table is 50 percent saturated.

Solution The analysis of this problem consists of the following steps:

1. Before lowering the G.W.T:
    The effective pressure at C is

$$(40)(130 - 62.4) + (20)(120 - 62.4) = 3856 \text{ lb/ft}^2$$

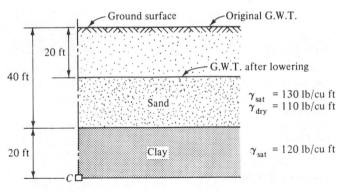

Figure P-4.13

2. After lowering the G.W.T:
    The density of wet sand above the G.W.T. is

$$\gamma_m = 110 + 0.5(130 - 110) = 120 \text{ lb/ft}^3$$

The effective pressure at C is

$$(20)(120) + (20)(30 - 62.4) + (20)(120 - 62.4) = 4904 \text{ lb/ft}^2$$

Thus the increase in effective pressure at C due to lowering of the G.W.T. is

$$\Delta p = 4904 - 3856 = 1048 \text{ lb/ft}^2$$

4.14 For the soil profile of Fig. P-4.14a, show the pressure diagram along a 20-ft-deep cut. Identify the effective pressure, the pore-water pressure, and the total pressure. Referring to Table 4-6, assume: $H_1 = 5$ ft, $H_2 = 15$ ft, $H = H_1 + H_2 = 20$ ft, $\gamma_w = 62.5$ lb/ft^2, $\gamma_{sat} = 125$ lb/ft^3, and $n = 0.5$.

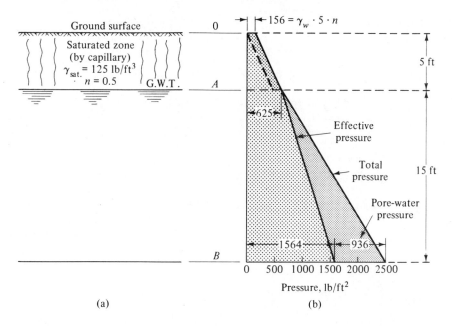

(a)   (b)

Figure P-4.14

By Table 4-6, the pressures are computed at three levels as shown below.   *Solution*

Pressure at Level *0*:

Capillary rise pressure	$H_1 \gamma_w n$	156 lb/ft²
Intergranular pressure	0	0
Total pressure		156 lb/ft²

Pressure at Level *A*:

Pore-water pressure	0	0
Intergranular pressure	$H_1 \gamma_{sat}$	625 lb/ft²
Total pressure		625 lb/ft²

Pressure at Level B:

Pore-water pressure	$H_2\gamma_w$	936 lb/ft²
Intergranular pressure	$H\gamma_{sat} - H_2\gamma_w$	1564 lb/ft²
Total pressure		2500 lb/ft²

Note that at the level B the total pressure is

$$p = H\gamma_{sat} = (20)(125) = 2500 \text{ lb/ft}^2$$

which is used to check the closing point of the pressure diagram in Fig. P-4.14b

SUPPLEMENTARY PROBLEMS

Static Equilibrium Method

4.15 Calculate the height of capillary rise in a glass tube that has an inside diameter of 0.1 in. Surface tension of water is 0.00504 lb/ft and angle of contact is assumed to be 0°.

Answer $h_c = 3.24$ in.

4.16 The effective grain size of a silty soil is $D_{10} = 0.007$ mm and its void ratio is $e = 0.50$. Using Hazen's formula (4-3), calculate the height h_c of capillary rise in this soil.

Answer h_c will vary between 286 and 1430 cm, depending on the value of C used.

Static Effects of Soil Moisture

4.17 Compute the effective pressure, the pore-water pressure, and the total pressure at A, B, and C of the soil profile shown in Fig. P-4.17.

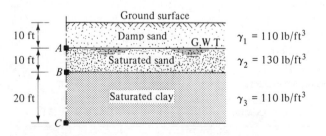

Figure P-4.17

Results computed by Table 4-6 are recorded in Table P-4.17.   *Answer*

TABLE P-4.17

Level	A	B	C
Effective pressure, in lb/ft²	1100	1776	2728
Pore-water pressure, in lb/ft²	0	624	1872
Total pressure, in lb/ft²	1100	2400	4600

4.18   The G.W.T. in a silty deposit was lowered from 5 ft to a depth of 15 ft. Calculate the increase in the effective pressure at a depth of 25 ft resulting from the lowering of the ground water table. Assume that soil above the G.W.T. will be saturated with capillary moisture before and after the lowering of the water table.

$p = 624 \, lb/ft^2$ *Answer*

4.19   For the soil profile of Fig. P-4.19, compute the pore-water pressure, the effective pressure, and the total pressure at point E.

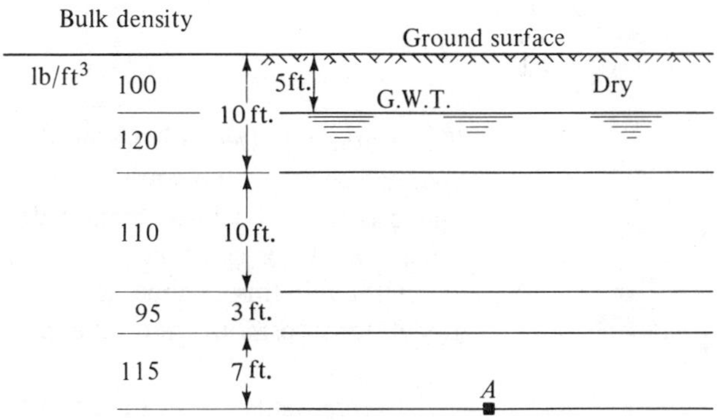

Figure P-4.19

By Table 4-5, $u_w = 1560$, $\bar{p} = 1730$, $p = 3290$, all in lb/ft².   *Answer*

soil
water
in
motion

5

5.1 INTRODUCTION

Concept

The stationary free soil water discussed in the preceding chapter is only a special case of the soil-water flow. Any change in the state of static equilibrium converts the soil cavities from water-storage reservoirs to passageways through which the water can flow.

Although this change may be induced by mechanical, thermal, gravitational, ionic, osmotic, and other gradients, the most common causes of flow are

1. *Water addition*, in the form of (a) rainfall, (b) inflow, or (c) both.
2. *Water loss*, in the form of (a) evaporation, (b) outflow, or (c) both.

The study of soil water in motion is one of the most important aspects of engineering soil mechanics, the summary of which is presented in this chapter.

Fluid Mechanics Classification

In general, the fluid flows fall into two major categories:

1. The *turbulent flow*, the fluid particles of which move in very irregular (chaotic) paths, causing losses of energy approximately proportional to the

square of flow velocity. The development of this flow requires a high velocity and large flow profiles.

2. The *laminar flow*, the fluid particles of which move in smooth, orderly streams (laminas), causing energy losses directly proportional to flow velocity. The development of this flow requires high viscosity, low velocity, and small flow profiles.

The turbulent flow rarely occurs in soils (except in shattered rock and coarse-gravel deposits). The laminar flow, however, is characteristic for all soils finer than coarse gravels, and, unless specifically stated otherwise, all water flows in this chapter are laminar.

Soil Mechanics Classification

Furthermore, soil mechanics divides the laminar flow into two particular systems:

1. The *nonsaturated flow*, the transfer mechanism of which depends on the degree of saturation (vapor transfer exceeds the liquid transfer or vice versa).
2. The *saturated flow*, the transfer mechanism of which is the liquid transfer through a fully saturated medium.

The study of the saturated flow is based on simple concepts and is introduced herein. The investigation of the nonsaturated flow is far more complex, and its presentation is beyond the scope of this volume.

Definitions

Several terms are frequently used in the analysis of flow: the steady and nonsteady flow, the uniform and nonuniform flow, the permeability, the seepage, and the drainage.

1. The *steady flow* is defined as a flow, the characteristics (velocity, density, pressure, etc.) of which are independent of time, temperature, and position. Once any or all of the flow characteristics become variables of any or all of the parameters mentioned above, the unsteady flow designation is used.
2. The *uniform flow* is defined as a flow, the velocity vector of which is identical in magnitude for any given instant at every point. Once the velocity vector becomes a function of the position, the designation of nonuniform flow is used.
3. The *permeability* (fluid conductivity) is the facility with which water and/ or other fluids are able to travel (permeate) through the soil.
4. The *seepage* is the process of transferring liquid through or in the soil (inflow).
5. The *drainage* is the process of transferring liquid from one location to another location in or out of the soil (outflow, relocation).

The last two terms are often used in a special sense, and their meaning must then be interpreted from the context.

Constants

In addition to the constants given in Section 4.1, the viscosity of distilled water $v_w = 10^{-5}$ g sec/cm² at room temperature (68 °F or 20 °C) is of practical importance. The values of v_w at various temperatures are given in Table 5-1. Other measures (pertinent to the study of soil water flow) and their conversion factors are recorded in Table 5-2.

TABLE 5-1   *Viscosity of Water (Values Are in Millipoises)*

°C	0	1	2	3	4	5	6	7	8	9
0	17.94	17.32	16.74	16.19	15.68	15.19	14.73	14.29	13.87	13.48
10	13.10	12.74	12.39	12.06	11.75	11.45	11.16	10.88	10.60	10.34
20	10.09	9.84	9.61	9.38	9.16	8.95	8.75	8.55	8.36	8.18
30	8.00	7.83	7.67	7.51	7.36	7.31	7.06	6.92	6.79	6.66
40	6.54	6.42	6.30	6.18	6.08	5.97	5.87	5.77	5.68	5.58
50	5.29	5.40	5.32	5.24	5.15	5.07	4.99	4.92	4.84	4.77
60	4.70	4.63	4.56	4.50	4.43	4.47	4.31	4.24	4.19	4.13
70	4.07	4.02	3.96	3.91	3.86	3.81	3.76	3.71	3.66	3.62
80	3.57	3.53	3.48	3.44	3.40	3.36	3.32	3.28	3.24	3.20
90	3.17	3.13	3.10	3.06	3.03	2.99	2.96	2.93	2.90	2.87
100	2.84	2.82	2.79	2.76	2.73	2.70	2.67	2.64	2.62	2.59

1 dyne sec/cm² = 1 poise
1 gram sec/cm² = 980.7 poises
1 pound sec/ft² = 478.69 poises
1 poise = 1000 millipoises

International Critical Tables, *Vol. V, McGraw-Hill, New York, 1929.*

TABLE 5-2   *Conversion Factors*

1. Area
    1 in.² = 6.4516 cm²   1 cm² = 0.1555 in.²
    1 ft² = 929.0 cm²   1 m² = 10.76 ft²
2. Volume
    1 in.³ = 16.39 cm³   1 cu cm = 0.06102 in.³
    1 ft³ = 28320 cm³   1 cu m = 35.31 ft³
3. Velocity
    1 in./sec = 2.540 cm/sec = .025 m/sec
    1 cm/sec = 0.3937 in./sec = .03281 ft/sec
    1 ft/sec = 30.48 cm/sec = 18.29 m/min
    1 m/sec = 3.281 ft/sec = 196.8 ft/min
    1 ft/min = 0.5080 cm/sec = 0.3048 m/min
    1 m/min = 0.05468 ft/sec = 3.281 ft/min

Conversion Factors for Engineers, *Dorr-Oliver, Inc., Stamford, Conn., 1961.*

5.2 PERMEABILITY

Characteristics

The *permeability* of soils, defined, before (Section 5.1, page 97) as the fluid conductivity, is one of the major soil characteristics affecting the rate of seepage of water under dams, drainage of subgrades, expulsion of water by heavy foundation loads, and so forth.

The *magnitude of permeability* is controlled by many factors, the most important of which are listed below.

1. The *grain size*. It appears that the permeability is proportional to the square of the effective grain size.
2. The *void ratio*. It appears that the permeability is proportional to the square of the void ratio.
3. The *fluid viscosity*. The permeability is inversely proportional to the fluid viscosity, which in turn is inversely proportional to the temperature.
4. The *structure of pores*. There is no analytical representation of this factor, but experiments indicate that the profile, connectivity, and smoothness of these miniature conduits are important.
5. The *degree of saturation*. Because the air or gases in voids reduce the flow profile (or may block some pores completely), the amount of initial moisture is an important factor.

Because of the indeterminacy of some of these factors, the determination of permeability is a constitutive problem, the governing equation of which must be determined empirically.

Poiseuille's Law

Like the static soil moisture, the flow of soil moisture may be visualized as a flow in a system of small tubes corresponding to soil pores. The experimental results related to the fluid flow in small glass tubes, first reported by G. H. Hagen (1839), verified independently by J. L. Poiseuille (1840), and analytically supported by R. K. Wiedeman (1856), led to the flow discharge equation, which is given as

$$Q = \frac{\gamma_w}{8v_w} r^2 i A \qquad\qquad (5\text{-}1)$$

where Q = total discharge
γ_w = unit weight of water
r = radius of one tube
i = hydraulic gradient
v_w = coefficient of viscosity of water
A = total flow area

This formula, which is know as *Poiseuille's law*, shows that the amount of discharge is directly proportional to the square of the tube radius.

Darcy's Laws

Poiseuille's law, if applied to the soil-water flow, implies that the discharge of two samples of soil of an identical r^2A must be the same value, even if one is composed of fine-grained and the other one of coarse-grained material. H. Darcy (1856), after several carefully executed experiments using the apparatus of Fig.

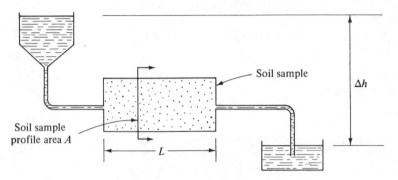

Soil sample

Δh

Soil sample profile area A

L

Figure 5-1   *Schematic diagram of Darcy's apparatus*

5-1, concluded that such is not the case and postulated two laws, which are still the basis for estimating the seepage of water in soils.

1. *Darcy's Law (laminar flow)*. For laminar flow in fully saturated soil of grain size less than 1.00 mm, the total discharge is

$$Q = vA = kiA \tag{5-2}$$

where v = discharge velocity,
       $i = \Delta h/L$ (as shown in Fig. 5-1) is the hydraulic gradient,
       k = permeability (coefficient of conductivity).

2. *Darcy's Law (turbulent flow)*. For turbulent flow in fully saturated soil of grain size more than 1.00 mm, the total discharge is

$$Q = vA = ki^\phi A \tag{5-3}$$

where ϕ is the exponent of turbulence (experimental results show $0.65 < \phi < 1.00$).

In these equations [(5-2) and (5-3)], the discharge velocity $v_d = v$ must equal the approach velocity $v_a = v$, but the seepage velocity v_f in soil computed from $Q = Av = A_v v_f = Anv_f$ is

$$v_f = \frac{v}{n}$$

where n is the porosity of soil layer and An is the area of void profile.

The quantities involved in (5-2) and (5-3) are given in the following units:

Q	k	A	v	i	ϕ
cm³/sec	cm/sec	cm²	cm/sec	Dimensionless	Dimensionless

Approximate Values of k

For granular soils, k may be estimated by A. Hazen's formula (Problem 5.1) as

$$k = C(D_{10})^2 \tag{5-4}$$

where $C = 100$ to 150 cm/sec is an empirical constant and D_{10} is the effective diameter (2-2) in percent.

For mixed-grained sandy soils, k may be estimated by A. Casagrande's formula (Problem 5-2) as

$$k = 1.4e^2(k_{0.85}) \tag{5-5}$$

where k is the permeability at the void ratio e (any void ratio) and $k_{0.85}$ is the permeability of the same soil at the void ratio of 0.85.

For a broad spectrum of soils, Table 5-3, known as Hough's Chart of Typical Permeability Coefficients, is available as an approximate information.

Laboratory Determination of k

The complexity of factors influencing soil permeability allows only a crude estimate of k by the methods stated above. Apparently the only reliable and rational method of finding the value of k is the laboratory test. Two standard tests are introduced here for this purpose.

1. The *constant-head permeameter test* gives reliable results for materials with high-fluid conductivity (coarse sands and gravel).
2. The *falling-head permeameter test* gives reliable results for soils with low fluid conductivity (fine sand, silt, and clay)

For additional information, such as the descriptions of the equipment and the procedures used in these tests, refer to Problems 5.3 and 5.4 respectively.

TABLE 5-3   *Chart of Typical Permeability Coefficients*

	PARTICLE-SIZE RANGE				"EFFECTIVE" SIZE		PERMEABILITY COEFFICIENT k		
	Inches		Millimeters						
	D_{max}	D_{min}	D_{max}	D_{min}	D_{20}(in.)	D_{10}(mm)	ft/year	ft/month	cm/sec
TURBULENT FLOW Derrick stone	120	36	–	–	48	–	100×10^6	100×10^5	100
One-man stone	12	4	–	–	6	–	30×10^6	30×15^5	30
Clean, fine to coarse gravel	3	$\frac{1}{4}$	80	10	$\frac{1}{2}$	–	10×10^6	10×10^5	10
Fine, uniform gravel	$\frac{3}{8}$	$\frac{1}{16}$	8	1.5	$\frac{1}{8}$	–	5×10^6	5×10^5	5
Very coarse, clean, uniform sand	$\frac{1}{8}$	$\frac{1}{32}$	3	0.8	$\frac{1}{16}$	–	3×10^6	3×10^5	3
LAMINAR FLOW Uniform, coarse sand	$\frac{1}{8}$	$\frac{1}{64}$	2	0.5	–	0.6	0.4×10^6	0.4×10^5	0.4
Uniform, medium sand	–	–	0.5	0.25	–	0.3	0.1×10^6	0.1×10^5	0.1
Clean, well-graded sand and gravel	–	–	10	0.05	–	0.1	0.01×10^6	0.01×10^5	0.01
Uniform, fine sand	–	–	0.25	0.05	–	0.06	4000	400	40×10^{-4}
Well-graded, silty sand and gravel	–	–	5	0.01	–	0.02	400	40	4×10^{-4}
Silty sand	–	–	2	0.005	–	0.01	100	10	10^{-4}
Uniform silt	–	–	0.05	0.005	–	0.006	50	5	0.5×10^{-4}
Sandy clay	–	–	1.0	0.001	–	0.002	5	0.5	0.05×10^{-4}
Silty clay	–	–	0.05	0.001	–	0.0015	1	0.1	0.01×10^{-4}
Clay (30 to 50 % clay sizes)	–	–	0.05	0.0005	–	0.0008	0.1	0.01	0.001×10^{-4}
Colloidal clay ($-2\mu \leq 50\%$)	–	–	0.01	10 Å	–	40 Å	0.001	10^{-4}	10^{-9}

K. B. Hough, Basic Soil Engineering, *Copyright 1957, Ronald Press, New York.*

5.3 SEEPAGE

Flow Net

Whereas the fluid conductivity of soil is called the permeability, the process of water transfer in soil (conduction) is called the *seepage*.

The graphical representation of the soil-water seepage is called the *flow lines*. Each flow line is the path followed by a particular drop of water and, for laminar flow, these lines never intersect. Similarly, the graphical representation of points of equal potential energy (head) is the *equipotential line*.

The steady-state flow of water through saturated soil can be represented by a system of flow lines and equipotential lines called the *flow net*, shown in Fig. 5-2.

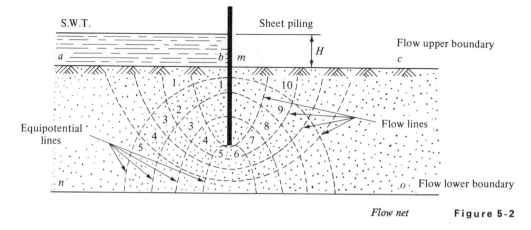

Flow net **Figure 5-2**

Laplace Equation

The mathematical representation of the flow net for the semi-infinite soil, the permeabilities of which in the x and z directions are k_x, k_z, respectively, is given by the *Laplace differential equation* (Problem 5.5) as

$$k_x \frac{\partial^2 h}{\partial x^2} + k_z \frac{\partial^2 h}{\partial z^2} = 0 \qquad\qquad (5\text{-}6)$$

where

$$\frac{\partial h}{\partial x} = i_x \quad \text{and} \quad \frac{\partial h}{\partial z} = i_z$$

are the *Darcy's law gradients* in the x and z directions, respectively.

The solution of this problem involves finding the $h = \psi(x, z)$ that satisfies both the differential equation and the boundary conditions associated with the problem.

Graphical Solution

The integral of (5-6) can be obtained by one of many methods available for the solution of the Laplace equation (complex variables, series, transforms, numerical methods, etc.). Because of the complexity involved in their solution, the approximate methods developed by P. Forchheimer and A. Casagrande (Problems 5.6 to 5.15) are frequently used.

Homogeneous Soils

The general form of (5-6) in the case of isotropic permeability, $k_x = k_z = k$, reduces to

$$k\frac{\partial^2 h}{\partial x^2} + k\frac{\partial^2 h}{\partial z^2} = 0 \qquad (5\text{-}7)$$

and the flow net becomes a systems of orthogonal curvilinear trajectories.

A further simplification takes place if the seepage proceeds through a pervious layer in the horizontal direction only. Then (5-7) reduces to

$$k\frac{\partial^2 h}{\partial x^2} = 0 \quad \text{and} \quad v = k\frac{\partial h}{\partial x} = ki_x \qquad (5\text{-}8)$$

where, by Fig. 5-3, $i_x = h/L$, and, by (5-2), $v = Q/A$.

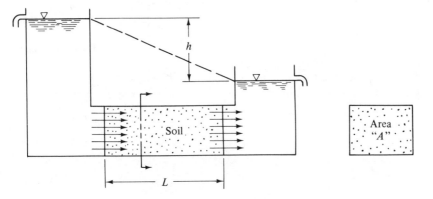

Figure 5-3   *Horizontal seepage through a pervious layer*

Finally, if this layer is inclined and the angle of inclination α is constant, (5-8) becomes

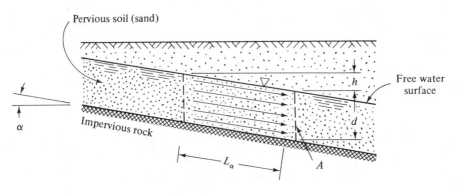

Figure 5-4   *Inclined seepage through a pervious layer*

$$v_\alpha = k\frac{h}{L_\alpha} = k\sin\alpha \qquad (5\text{-}9)$$

where L_α is measured along the sloped plane of Fig. 5-4 and $v_\alpha = Q/A = q/d$ with d representing the vertical depth of the layer and q standing for the flow quantity per unit width of the profile A (Problem 5.16).

Stratified Soils

For soils of Fig. 5-5 consisting of m horizontal layers of thickness $L_1, L_2, \ldots,$

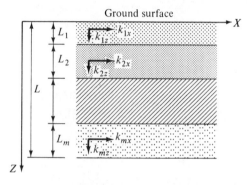

Seepage in stratified soil   **Figure 5-5**

L_m and permeabilities $k_{1z}, k_{2z}, \ldots, k_{mz}$, the equivalent permeability of the whole system in the vertical direction is

$$k_z = \frac{L_1 + L_2 + \cdots L_m}{L_1/k_{1z} + L_2/k_{2z} + \cdots + L_m/k_{mz}} = \frac{L}{\lambda_z} \qquad (5\text{-}10)$$

and the discharge in the vertical direction per unit area normal to the flow direction is

$$q_z = k_z\frac{h}{L} = \frac{h}{\lambda_z} \qquad (5\text{-}11)$$

Similarly, the equivalent permeability of the same system in the horizontal direction is

$$k_x = \frac{L_1 k_{1x} + L_2 k_{2x} + \cdots + L_m k_{mx}}{L_1 + L_2 + \cdots + L_m} = \frac{\lambda_x}{L} \qquad (5\text{-}12)$$

and the discharge in the horizontal direction per unit area normal to the flow direction is

$$q_x = k_x \frac{h}{L} = \frac{h\lambda_x}{L^2} \tag{5-13}$$

The derivations and applications of (5-10), (5-11) and (5-12), (5-13) are shown in Problems 5.17 to 5.21.

5.4 PRESSURE EFFECTS OF MOISTURE FLOW

General

The seepage of water through soil induces changes in the static pressures (intergranular and hydrostatic), discussed in Section 4.4, which in turn may affect the stability and strength of the soil. The typical cases of such effects are considered in this section.

Downward Flow

In the system of Fig. 5-6, with the downward flow of $i = h/H$, the intergranular pressure at B is increased by

$$\bar{p}_f = iH\gamma_w = h\gamma_w \tag{5-14}$$

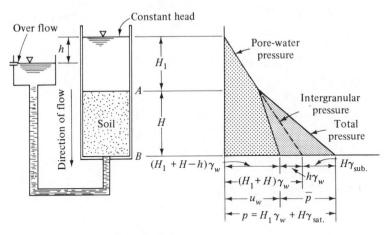

Figure 5-6   *Downward flow*

called the seepage pressure, produced as a result of the downward drag of the moving fluid (Problem 5.22).

Upward Flow

In the system of Fig. 5-7, with the upward flow of $i = h/H$, the intergranular pressure at B is reduced by $\bar{p}_f$ of (5-14), produced as a result of the upward drag of the moving fluid (Problem 5.22).

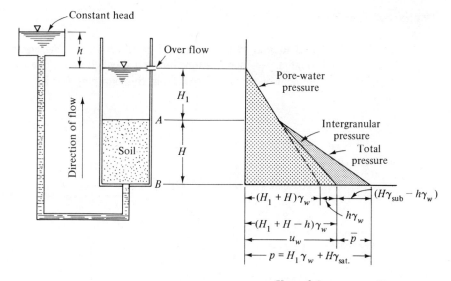

Upward flow

Figure 5-7

Critical Hydraulic Gradient

In the latter case, the seepage pressure may be increased by increasing the hydraulic gradient until the intergranular pressure becomes zero. In such a case,

$$H\gamma_{\text{sub}} = i_c H\gamma_w \quad \text{and} \quad i_c = \frac{\gamma_{\text{sub}}}{\gamma_w} \tag{5-15}$$

where i_c is the critical hydraulic gradient at which the intergranular pressure becomes zero. This is very important in foundation design problems because, under such a condition, cohesionless soils become very loose and cannot carry loads. Most of the boiling or quicksand conditions are due to this critical condition (Problems 5.23 to 5.26).

5.5 FIELD PERMEABILITY TESTS

General

The most reliable determination of the coefficient of permeability k is obtained by conducting a pumping test in the field. This is particularly true for coarse-grained material below the ground water table.

The field tests are relatively expensive and are considered only in cases where the question of permeability seems of major importance. In such cases, a star system of wells is drilled with the major pumping well placed at the center and several satellite observation wells placed at some distance from this center.

Ordinary Perfect Wells

If a field pumping test (Fig. 5-8) is conducted by sinking a pumping well through a pervious stratum, the well is called an *ordinary perfect well* and the coefficient of permeability k from this pumping test is

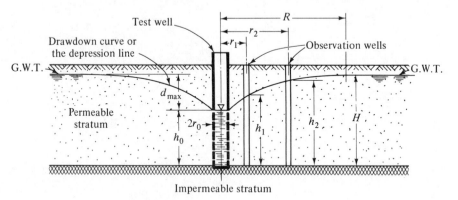

Figure 5-8   *Ordinary perfect wells*

$$k = \frac{Q \ln (r_2/r_1)}{\pi(h_2^2 - h_1^2)} \quad \text{or} \quad k = \frac{2.3Q \log (r_2/r_1)}{\pi(h_2^2 - h_1^2)} \tag{5-16}$$

where Q = rate of flow into the well,

     r_1, r_2 = radial distances to the center line of the well,

     h_1, h_2 = heights of water table at radial distances r_1 and r_2

According to J. S. Kozeny (1933), the maximum radius of influence R is

$$R = \sqrt{\frac{12t}{n} \sqrt{\frac{Qk}{\pi}}} \tag{5-17}$$

where t is the time during which a discharge of Q from the well has been established, and k and n are the coefficients of permeability and porosity respectively.

The maximum yield capacity was derived by J. Dupuit (1863) and is expressed as follows:

$$Q = \frac{\pi k(H^2 - h_0^2)}{2.3 \log (R/r_0)} \quad \text{or} \quad Q = \frac{\pi k(2H - d_{\max}) d_{\max}}{2.3 \log (R/r_0)} \tag{5-18}$$

where h_0 = height of water in well after pumping,

r_0 = radius of well,

$d_{\max}$ = maximum drawdown (distance from original G. W. T.),

H = elevation of G. W. T. above bottom of well (Fig. 5-8)

For any point on the depression curve, the coordinates x and y represent the radial distance and the elevation of the point above the base of well.

The equation of the depression line, given as

$$y = \sqrt{H^2 - \left[\frac{Q}{\pi k}(2.3) \log\left(\frac{R}{x}\right)\right]} \qquad (5\text{-}19)$$

takes two special forms.

For $x = r_0, y = h_0$,

$$h_0 = \sqrt{H^2 - \frac{2.3Q}{\pi k}\left[\log\left(\frac{R}{r_0}\right)\right]} \qquad (5\text{-}20a)$$

For $x = R$, $\quad \log\left(\frac{R}{R}\right) = 0$, and

$$y = H \qquad (5\text{-}20b)$$

The application of (5-16) to (5-20) is shown in Problems 5.27 to 5.32.

Perfect Artesian Wells

When a pumping well penetrates a confined aquifer (Fig. 5-9), the well is called a *perfect* or *fully penetrating artesian well*. In this case, the coefficient

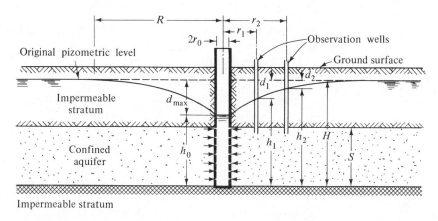

Perfect artesian wells   **Figure 5-9**

of permeability k from a field pumping test is

$$k = \frac{Q}{2.73S} \left[\frac{\log (r_2/r_1)}{d_1 - d_2} \right] \tag{5-21}$$

where S is the thickness of water-bearing stratum, called *aquifer*.
The rate of flow into the well is

$$Q = 2.73 \frac{kSd_{\max}}{\log (R/r_0)} \tag{5-22}$$

and, for any point on the depression curve, the elevation y above the base of the well is

$$y = h_0 + \frac{Q}{2.73kS} \log \left(\frac{x}{r_0} \right) \tag{5-23}$$

The applications of these equations are given in Problems 5.33 and 5.34.

SOLVED PROBLEMS

Permeability

5.1 Estimate the coefficient of permeability k of a sand filter that has an effective size diameter D_{10} of 0.065 mm.

Solution Since the soil is a sandy material, use Hazen's approximate formula (5-4) and assume a value of $C = 100$.
Thus the coefficient of permeability

$$k = C(D_{10})^2 = 100(0.0065)^2 = 0.0042 \text{ cm/sec}$$

5.2 Estimate the coefficient of permeability k of a mixed-grained sandy soil at a void ratio of 0.78. The same soil has a permeability coefficient of $(2)10^{-2}$ cm/sec at a void ratio of 0.66.

Solution By (5-5),

$$k_{0.66} = 1.4e^2 k_{0.85}$$

from which, in terms of $k_{0.66} = (2)10^{-2}$ cm/sec and $e = 0.66$,

$$k_{0.85} = \frac{k_{0.66}}{1.4e^2} = \frac{(2)10^{-2}}{(1.4)(0.66)^2} = (3.28)10^{-2} \text{ cm/sec}$$

Similarly, the permeability coefficient at void ratio $e = 0.78$ is

$$k_{0.78} = 1.4(0.78)^2(3.28)(10^{-2}) = (2.79)10^{-2} \text{ cm/sec}$$

5.3   Describe the procedure for laboratory determination of the permeability of soils, using the constant-head permeameter test.

The test arrangement is illustrated in Fig. P-5.3. The sample is placed in a cylindrical   *Answer* container of length L and cross-sectional area A. Water flows through the sample under a head difference h, which is kept constant. The amount of water flowing through the sample Q during a time interval t is determined by weighing the water collected in the cup at the bottom of the sample. The permeability k is then determined from the relationship

$$k = \frac{QL}{Aht}$$

This test gives more reliable results for high-permeability soil-clean sands and fine gravels, for example.

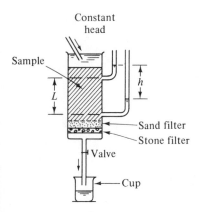

Figure P-5.3

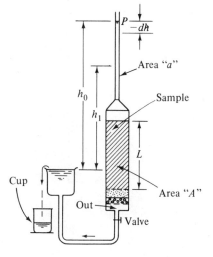

Figure P-5.4

5.4   Describe the procedure for the laboratory determination of the permeability of soils by the falling-head permeameter test.

The arrangement for the falling-head permeameter test is illustrated in Fig. P-5.4.   *Solution*
It can be seen that the arrangement is somewhat similar to the constant-head permeameter, for a cylindrical sample of soil is placed in a container of length L and cross-sectional area A. Water flows through the sample under an initial head difference of h_0. The water level in the vertical tube, with a cross-sectional area a, drops as flow progresses; and at the end of the time interval t, the head difference is h_1.
Since the head varies throughout the duration of the test, the rate of flow during this interval becomes

$$\frac{\partial Q}{\partial t} = -\frac{a\,\partial h}{\partial t} = k\left(\frac{h}{L}\right)A$$

Then, integrating between limits of h_0 and h_1 for h and t_0 and t_1 for t,

$$k = \frac{aL}{At}\ln\frac{h_0}{h_1} \quad \text{or} \quad 2.3\frac{aL}{At}\log\frac{h_0}{h_1} \qquad \text{(a)}$$

The falling-head permeameter is more suitable for tests on materials of low permeability, such as clays and silty clays.

Seepage

5.5   Derive the Laplace differential equation (5-4), representing mathematically the flow net in a semi-infinite, saturated soil space. Assume the permeabilities k_x, k_y, k_z in the x, y, z directions to be known. Furthermore, assume a laminar steady-state flow of incompressible water of γ_w at constant temperature for which Darcy's law (1), equation (5-2), page 100, is valid.

Solution   For the elemental soil cube $dx\,dy\,dz$ of Fig. P-5.5, the inflow and outflow velocities in the x, y, z directions are given by (5-2) as shown below.

The inflow velocities (near side of the cube) are

$$v_x = k_x i_x = \frac{k_x\,\partial h}{\partial x}, \quad v_y = k_y i_y = \frac{k_y\,\partial h}{\partial y},$$

$$v_z = k_z i_z = \frac{k_z\,\partial h}{\partial t} \qquad \text{(a)}$$

The outflow velocities (far side of the cube) are

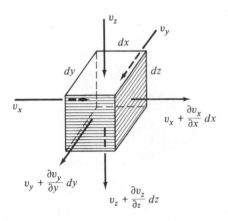

$$v_x + \frac{\partial v_x}{\partial x}\,dx = k_x\left(\frac{\partial h}{\partial x} + \frac{\partial^2 h}{\partial x^2}\,dx\right),$$

$$v_y + \frac{\partial v_y}{\partial y}\,dy = k_y\left(\frac{\partial h}{\partial y} + \frac{\partial^2 h}{\partial y^2}\,dy\right),$$

$$v_z + \frac{\partial v_z}{\partial z}\,d_z = k_z\left(\frac{\partial h}{\partial z} + \frac{\partial^2 h}{\partial z^2}\,dz\right) \qquad \text{(b)}$$

Figure P-5.5

In both cases, h is the hydraulic head at (x, y, z), and the assumption of perfect continuity and small changes is valid.

With the velocities [known, the rate of storage of water in the cube is given as the difference between the mass inflow and outflow flux, recorded in Table P-5.5 and designated as Δm.

In terms of Table P-5.5 and equations (a), (b),

$$\Delta m = -G_w\left(k_x\frac{\partial^2 h}{\partial x^2} + k_y\frac{\partial^2 h}{\partial y^2} + k_z\frac{\partial^2 h}{\partial z^2}\right) \qquad \text{(c)}$$

For a steady-state flow, $\Delta m = 0$, and (c) reduces to

Mass Inflow Flux	Mass Outflow Flux
$G_w v_x \, dy \, dz$	$G_w \left(v_x + \dfrac{\partial v_x}{\partial x} dx \right) dy \, dz$
$G_w v_y \, dz \, dx$	$G_w \left(v_y + \dfrac{\partial v_y}{\partial y} dy \right) dz \, dx$
$G_w v_z \, dx \, dy$	$G_w \left(v_z + \dfrac{\partial v_z}{\partial z} dz \right) dx \, dy$

NOTE: G_w = specific mass of water.

$$k_x \frac{\partial^2 h}{\partial x^2} + k_y \frac{\partial^2 h}{\partial y^2} + k_z \frac{\partial^2 h}{\partial z^2} = 0 \qquad \text{(d)}$$

For a two-dimensional flow, (d) becomes

$$k_x \frac{\partial^2 h}{\partial x^2} + k_z \frac{\partial^2 h}{\partial z^2} = 0 \qquad \text{(e)}$$

which is the desired result (5-6).

5.6 Outline the procedure for drawing the flow net for two-dimensional steady flow in a saturated soil for the sheet pile illustrated in Fig. 5-2.

The procedure is as follows: *Solution*

1. First define the flow boundaries. Generally there are four boundary conditions:
   - (a) the entrance surface ⎤
   - (b) the exit surface   ⎦ equipotential lines
   - (c) the upper flow line ⎤
   - (d) the lower flow line ⎦ flow lines

   These lines are shown in Fig. 5-2, where *ab* and *mc* are upstream and downstream equipotentials, respectively, the upper flow line is *bhm* (along the surfaces of the sheet pile), and the lower flow line is the surface of the impervious layer *no*.

2. Within these boundaries, the flow net of Fig. 5-2 can be drawn according to the following rules:
   - (a) Flow lines should be drawn approximately parallel to each other and should never cross one another.
   - (b) Equipotential lines are drawn such that they intersect the flow lines at right angles.
   - (c) The equipotential lines are selected in such a way as to form an orthogonal net with the flow lines or approximate "squares."

In general, considerable training and experience are required before a student can draw flow nets accurately and quickly for different types of situations.

5.7 What are the practical uses of the flow nets?

Answer   Flow nets are used for

1. Determination of the rate of flow of water from and under water-retaining structures (e.g., reservoirs, dams).
2. Determination of the seepage pressure in the soil mass under water-retaining structures.

5.8   Using the flow net drawn in Fig. 5-2, show how you can determine the rate of flow under the sheet pile. Assume that the length of the sheet pile is 100 ft, the depth of water on one side of the sheet is 30 ft, and the other side is dry. Assume that the coefficient of permeability of the soil, $k = (20)10^{-4}$ cm/sec.

Solution   In Fig. 5-2, assume that

The number of flow lanes or paths between flow lines $= N_f$

The number of equipotential increments or drops $= N_e$

Now, consider a series of squares between any two flow lines (Fig. P-5.8). The equipotential lines in this figure represent pizometric pressure heads h_1, h_2, h_3, and h_4.

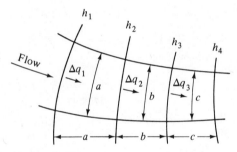

Figure P-5.8

Assuming that no flow takes places across the flow lines, then

$$\Delta q_1 = \Delta q_2 = \Delta q_3 = \Delta q \qquad \text{(a)}$$

From Darcy's law (5-2), the discharge Δq is

$$q = kiA \qquad \text{(b)}$$

Substituting (b) into (a) produces

$$k\left(\frac{h_1 - h_2}{a}\right)a = k\left(\frac{h_2 - h_3}{b}\right)b = k\left(\frac{h_3 - h_4}{c}\right)c$$

which is further reduced to

$$h_1 - h_2 = h_2 - h_3 = h_3 - h_4 = \Delta h = \frac{H}{N_e} \tag{c}$$

where H is the total pressure head and N_e is defined above.
 Thus the total rate of discharge Q in the entire flow net is

$$Q = (N_f)(\Delta q) = (N_f)(k)(\Delta h) = (k)(H)\left(\frac{N_f}{N_e}\right) \tag{d}$$

and N_f and N_e are equal to 5 and 10 respectively (Fig. 5-2).
 The head difference between two successive equipotential lines is therefore $H/10$.
 Then (d) in terms of given values of k, H, and N_f/N_e produces

$$Q = \frac{(20)10^{-4}}{30.5}(60)(30)\left(\frac{5}{10}\right)(100) = 5.89 \text{ ft}^3/\text{min}/100\text{-ft length of the sheet pile.}$$

5.9 For the earth dam with the reservoir level at the water table (W. T.) and no tail water, as shown in Fig. P-5.9a, determine the position of the discharge point of the line of seepage. Use the Casagrande graphical method.

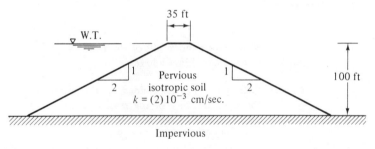

Figure P-5.9a

The procedure for determining the discharge point of the line of seepage using *Solution* Casagrande's method is illustrated in Fig. P-5.9b, and the steps are as follows;

1. Determine point b at $\frac{1}{3}$ the distance S.
2. Extend a vertical line from b to intersect the downstream slope of the dam (or its extension) at point c.
3. On oc as a diameter, draw a semicircle.
4. Find point d on the downstream slope of the dam by intersecting it with a line drawn parallel to the base of the dam and at the level of the water surface in the reservoir.
5. With point o as the center, and with the radius of od, draw arc dh.
6. From point c as a center and with a radius ch, draw arc he.
7. The intersection of arc he with the downstream face of the dam gives the outcrop of point e of the seepage line, and distance oe or a is measured and is $\cong 89.0$ ft.

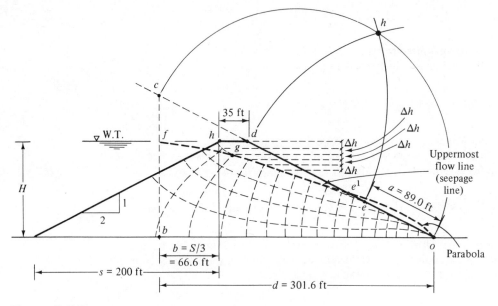

Figure P-5.9b

5.10   Using the results of Problem 5.9, plot the entire seepage line and the flow net to a scale of 1 in. = 50 ft.

Solution   After point *f* is determined on the water surface at a distance equal to *S*/3 from the upstream face of the reservoir, a parabolic curve with point *o* as the focus is drawn, starting from point *f*. The parabolic curve represents the theoretical uppermost flow line or the seepage line.

The actual entrance condition is then corrected by sketching in the arc *gh'* normal to the upstream slope and tangent to the parabolic free surface. The parabola will intersect the downstream face at point *e'*. The actual exit condition will also have to be corrected by sketching in an arc that is a tangent to the downstream face of the dam at point *e*, which was previously determined in Problem 5.7, point (7).

Having plotted the basic parabola and corrected the exit and entrance conditions, it is then an easy matter to draw, with a fair degree of approximation, the entire line of seepage.

Various conditions for entrance and exit for the line of seepage are given in Fig. P-5.10.

In order to draw the flow net, a convenient number of flow channels are selected (in this case, four channels), and the flow lines are drawn. Then the curve *hge* is divided into a convenient number of equal vertical intervals Δh and the flow net is sketched, following the guidelines of Problem 5.6. The final results are shown in Fig. P-5.9b.

5.11   Using the results of Problems 5.9, 5.10, and Fig. P-5.9b, compute the quantity of seepage water through the dam.

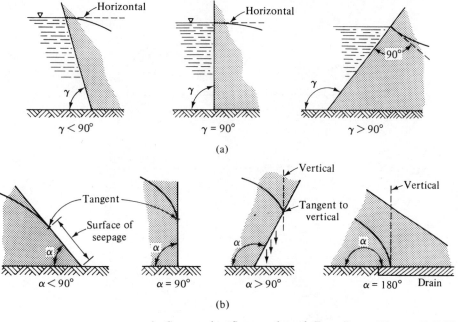

A. Casagrande, "Seepage through Dams," *Journal of the New England Water Works Association*, Vol. 1, No. 2, June 1937, p. 139

Figure P-5.10

By Fig. P-5.9b, the seepage quantity, given as (d) in Problem 5.8, is here   *Solution*

$$(q) = (k)(H)\left(\frac{N_f}{N_e}\right) = \frac{(2)(10^{-3})(60)}{30.5} (100) \frac{4}{17}$$

$$= (93.0)10^{-3} \text{ ft}^3/\text{min}$$

5.12   For the dam of Problem 5.9, compute the quantity q of seepage water through the dam in cubic feet per minute per unit length of this dam. Use the formulas and charts of Fig. P-5.12.

From Fig. P-5.9a and b,   *Solution*

$$\alpha = \tan^{-1} \frac{1}{2}$$

$$d \text{ (Fig. P-5.9b)} = 200 + 35 + 66.6 = 301.6 \text{ ft}$$

$$\frac{d}{H} = \frac{301.6}{100} = 3.0$$

With these values known, the quantity of seepage per unit length of the dam is, by Fig. P-5.12,

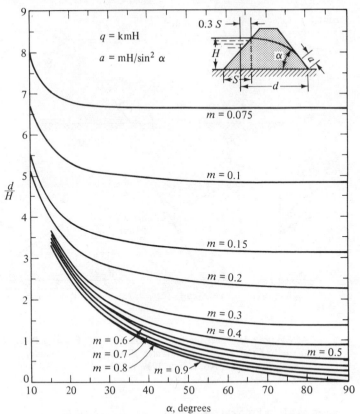

Figure P-5.12   G. Gilboy, "Hydraulic Fill Dams," *Proceedings
Intern. Comm. Large Dams*, Stockholm, 1933

$$q = kmH = \frac{(2)(10^{-3})(60)}{30.5}(0.2)(100) = (79)10^{-3} \text{ ft}^3/\text{min}$$

where $m = 0.2$ was located on the chart at the intersection of α and d/H. The
solution for q in Problem 5.11 was given as $(93)(10^{-3})$ ft^3/min.

The exit of the line of seepage (emergence) on the downstream side of the dam
(Fig. P-5.12) is at

$$a = \frac{mH}{\sin^2 \alpha} = \frac{(0.2)(100)}{(0.446)^2} = 100 \text{ ft}$$

whereas the graphical solution in Problem 5.9 gave $a = 89.0$ ft.

5.13 A gravel filter is placed horizontally under the toe of the earth dam in Fig.
P-5.9a. Show the effect of this filter on the seepage line.

Solution The pervious gravel filter blanket, placed as shown in Fig. P-5.13, with much
higher permeability than the material of the earth dam, provides a horizontal discharge surface.

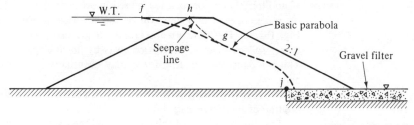

Figure P-5.13

A parabola is then drawn through point f as in Fig. P-5.9, but here the parabola has the focus at j, as shown in Fig. P-5.13. The entrance point will be corrected as in Fig. P-5.9b by drawing the arc gh. The exit point, however, will not need any correction in this case, for it intersects the horizontal discharge surface at a 90 degree angle (Fig. P-5.10b).

5.14 For the earth dam of Fig. P-5.9, a pervious rock-toe is placed as shown in Fig. P-5.14. Sketch the flow net and from it calculate seepage loss through the dam in cubic feet per minute per unit length of the dam.

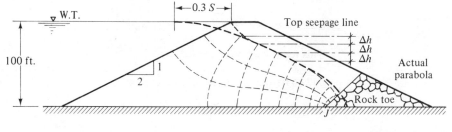

Figure P-5.14

From the flow net sketched in Fig. P-5.14, the number of flow channels, $N_f = 2.7$, the number of equipotential drops, $N_e = 9$, and the shape factor given as the ratio of these two coefficients become   *Solution*

$$\frac{N_f}{N_e} = \frac{2.7}{9} = 0.30$$

Then by (d) in Problem 5.8,

$$q = (k)(H)\left(\frac{N_f}{N_e}\right) = (2)(10^{-3})\left(\frac{60}{30.5}\right)(100)(0.30) = (118)10^{-3} \text{ ft}^3/\text{min}$$

where 60/30.5 is the conversion factor from centimeters per second to feet per minute.

5.15 Explain the purpose of placing a horizontal filter blanket or rock-toe in the earth dams of Problems 5.13 and 5.14 respectively.

Answer   If no filter blanket or rock-toe is used in an earth dam, such as that in Problem 5.9, the top seepage or flow line exits on the downstream slope of the dam, as shown in Fig. P-5.9b. The downstream face then gradually erodes, and the seepage force causes soil particles to be carried away by flowing water from the face of the dam. This is a dangerous process, for it may eventually lead to the failure of the entire dam.

In order to prevent such a failure, it is necessary to provide drains that lower the position of the top seepage line.

The filter blanket in Problem 5.13 and the rock-toe in Problem 5.14 are two possible forms of drains. Note how the top flow line in both cases is deflected downward and away from the downstream face of the dam.

5.16   An inclined flow similar to that in Fig. 5-4 takes place over an impervious stratum of rock through a fine-sand-sediment layer. The angle of inclination of the stratum is 3 degrees, and the depth of the sheet of flow is 5 ft. The coefficient of permeability of the sand layer is $k = 0.10$ ft/min. Compute the rate of discharge through a section 1.0 ft long, perpendicular to the direction of flow.

Solution   The rate of required discharge is given by (5-9) as

$$q = v_\alpha A = k \sin \alpha A = 0.02615 \ \text{ft}^3/\text{min per 1 ft of the sheet flow}$$

where

$$\sin \alpha = \sin 3° = 0.0523, \quad k = 0.10 \ \text{ft/min}, \quad A = (5)(1) = 5 \ \text{ft}^2$$

This result may be converted to

$$q = 37.66 \ \text{ft}^3/\text{day per 1 ft of the sheet flow}$$

which gives a better evaluation of the seepage discharge.

Stratified Soils

5.17   For the stratified deposit of Fig. 5-5, shown in the theory section of this chapter, derive the analytical expression (5-10) for the equivalent permeability of the system in the vertical direction.

Solution   Based on the principle of continuity, the rate of flow and its velocity through each stratum must be the same.

Thus

$$q_z = \frac{k_z h}{L} = \frac{k_{1z} h_1}{L_1} = \frac{k_{2z} h_2}{L_2} = \cdots = \frac{k_{mz} h_m}{L_m} \tag{a}$$

Also, the total head loss must be equal to the sum of all head losses of the respective layers; that is,

$$h = h_1 + h_2 + \cdots + h_m \tag{b}$$

where by (a)

$$h_1 = \frac{k_z h/L}{k_{1z}/L_1} \qquad h_2 = \frac{k_z h/L}{k_{2z}/L_1}, \ldots, h_m = \frac{k_z h/L}{k_{mz}/L_m} \tag{c}$$

With the substitution (c), (b) becomes

$$h = \frac{k_z h}{L} \left(\frac{L_1}{k_{1z}} + \frac{L_z}{k_{2z}} + \cdots + \frac{L_m}{k_{mz}} \right)$$

from which

$$k_z = \frac{L}{L_1/k_{1z} + L_2/k_{2z} + \cdots + L_m/k_{mz}} = \frac{L}{\lambda_z} \tag{d}$$

is the desired equivalent permeability in the vertical direction (5-10).

Consequently, the discharge in the vertical direction per unit area normal to the flow direction is then

$$q_z = \frac{k_z h}{L} = \frac{h}{\displaystyle\sum_{j=1}^{m} L_j/k_{jz}} = \frac{h}{\lambda_z} \tag{e}$$

which is (5-11).

These results show that the permeability in the vertical direction k_z is controlled by the least-permeable layer and may be always approximated as

$$k_z = \frac{L}{L_l/k_{lz}} \tag{f}$$

where L_l and k_{lz} are the thickness and permeability of the least-permeable layer.

5.18   The stratified deposit of Fig. P-5.18 has been selected as a dam site location. Using the results of Problem 5.17, compute the equivalent vertical permeability k_z.

Ground surface

$K_1 = 5 \times 10^{-3}$ ft/sec $L_1 = 5$ ft
$K_2 = 2 \times 10^{-3}$ ft/sec $L_2 = 8$ ft
$K_3 = 1 \times 10^{-5}$ ft/sec $L_3 = 4$ ft
$K_4 = 2 \times 10^{-7}$ ft/sec $L_4 = 10$ ft

k_x

k_z

$L = 27$ ft

Figure P-5.18

By (5-10), given as (d) in Problem 5.17,   *Solution*

$$k_z = \frac{27}{(5/5)10^3 + (8/2)10^3 + (4/1)10^5 + (10/2)10^7} = (5.36)10^{-7} \text{ ft/sec}$$

and by (f) of Problem 5.17,

$$k_z = \frac{27}{(10/2)10^7} = (5.4)10^{-7} \text{ ft/sec}$$

which shows the closeness of this approximation.

5.19 For the stratified deposit of Fig. 5-5, shown in the theory section of this chapter, derive the analytical expression (5-12) for equivalent permeability of the system in the horizontal direction.

Solution For the flow parallel to the stratification (bedding planes), the flow discharge equals the sum of partial discharges, each one of which corresponds to one layer; that is,

$$Q_x = Q_{1x} + Q_{2x} + \cdots + Q_{mx} \qquad (a)$$

where by (5-2)

$$Q_x = k_x i_x \underbrace{(L)(1)}_{A_x} \qquad (b)$$

$$Q_{1x}^{\rightarrow} = k_{1x} i_{1x} \underbrace{(L_1)(1)}_{A_{1x}} \qquad Q_{2x} = k_{2x} i_{2x} \underbrace{(L_2)(1)}_{A_{2x}}, \ldots, Q_{mx} = k_{mx} i_{mx} (L_m)(1) \qquad (c)$$

$$i_x = i_{1x} = i_{2x} = \cdots = i_{mx}$$

Then (a) in terms of (b) and (c) reduces to

$$k_x = \frac{k_{1x} L_1 + k_{2x} L_2 + \cdots + k_{mx} L_m}{L} = \frac{\lambda_x}{L} \qquad (d)$$

which is the desired equivalent permeability in the horizontal direction (5-12).

Consequently, the discharge in the horizontal direction per unit area normal to the flow direction is

$$q_x = \frac{k_x h}{L} = \frac{h \left(\sum\limits_{j=1}^{m} k_{jx} L_j \right)}{L^2} = \frac{h \lambda_x}{L^2} \qquad (e)$$

which is (5-13).

These results show that the permeability in the horizontal direction k_x is controlled by the permeability of the most permeable layer. If $k_{1x}, k_{2x}, \ldots, k_{mx}$ vary within a wide range, k_x may be approximated by

$$k_x \cong \frac{k_{kx} L_k}{L} \qquad (f)$$

where L_k and k_{kx} are the thickness and permeability of the most permeable layer. Here the expression (f), however, is a much poorer approximation than (f) in the preceding problem and is of a very limited value (see Problem 5.20).

5.20 Using the results of Problem 5.19, compute the equivalent horizontal permeability k_x for the stratified deposit of Fig. P-5.18.

By (5-12), given as (d) in Problem 5.19,   *Solution*

$$k_x = \frac{(5)(10^{-3})(5) + (2)(10^{-3})(8) + (1)(10^{-5})(4) + (2)(10^{-7})(10)}{27} = (1.52)10^{-3} \text{ ft/sec}$$

and by (f) in Problem 5.19,

$$k_x = \frac{(5)(10^{-3})(5)}{27} = (0.92)10^{-3} \text{ ft/sec}$$

which shows the inadequacy of this approximation.

5.21   A clay deposit 6.1 ft thick was found to have a layer of fine sand in the middle of its depth; the sand has a thickness of about 1.2 in. If the permeability of the sandy soil is $(5)10^{-4}$ ft/sec and that of the clay is $(5)10^{-6}$ ft/sec, calculate the ratio of the coefficient of horizontal and vertical permeabilities.

By (5-10), in terms of   *Solution*

$$L_1 = L_3 = 3 \text{ ft}, \quad L_2 = 0.1 \text{ ft}, \quad L = 6.1 \text{ ft},$$

and

$$k_{1z} = k_{3z} = \frac{k_{2z}}{100} = (5)10^{-6} \text{ ft/sec},$$

the vertical equivalent permeability is

$$k_z = \frac{L}{L_1/k_{1z} + L_2/k_{2z} + L_3/k_{3z}} = (5.08)10^{-6} \text{ ft/sec}$$

By (5-12), in terms of the same values as above and under the assumption of

$$k_{1x} = k_{1z}, \quad k_{2x} = k_{2z}, \quad k_{3x} = k_{3z},$$

the horizontal equivalent permeability is

$$k_x = \frac{k_{1x}L_1 + k_{2x}L_2 + k_{3x}L_3}{L} = (13.1)10^{-6} \text{ ft/sec}$$

The ratio $k_x/k_z = 2.58$ shows that although the permeability of each layer is the same number in the x and z directions, the horizontal permeability of the whole system equals 2.58 of the vertical permeability of the same system.

The lowering of the system permeability in the vertical direction is caused by the layer of clay, the thickness of which is insignificant compared to the depth of the system (0.1 versus 6.1 ft, less than 2 percent).

Pressure Effects of Moisture Flow

5.22   Two samples of soils were placed in two cylinders, as shown in Fig. P-5.22a and b. In (a) there is no outflow at the bottom of the cylinder. The other cylinder,

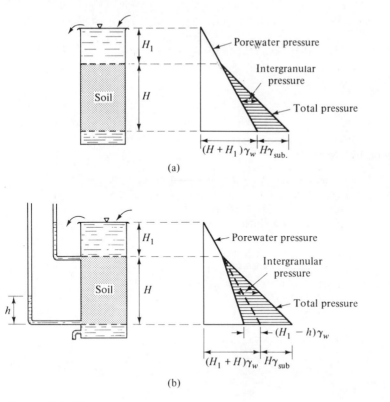

(a)

(b)

Figure P-5.22

however, is open from the bottom and water can pass through the soil sample down-ward. Draw the pore, intergranular, and total pressure diagrams for each case.

Solution   Comparing the pressure diagrams in Fig. P-5.22a and b, it is evident that the intergranular pressure in part (b) is higher than that in (a) by the value of $(H_1 - h)\gamma_w$. This difference results from the frictional drag of the flowing water on the soil particles, sometimes referred to as "seepage pressure."

When this pressure is in the same direction as that of the gravitational force of the soil, as in this problem, it increases the intergranular pressure (i.e., increases the stability) and is of little concern to the engineer. However, if the flow is against the weight or gravitational force of the soil, the intergranular pressure is reduced, and there is a tendency toward an unstable condition (see Problem 5.23).

5.23   A large, open excavation was made in a layer of clay with a saturated unit weight of 112 lb/ft.[3] Exploration of the site before excavation showed a stratum of dense sand at a depth of 40 ft below the ground surface. It was observed that the water had risen to an elevation of 15 ft below ground surface in one of the explora-tion drill holes. Calculate the critical depth of the excavation after which the bottom would be cracked and a boiling condition would exist (Fig. P-5.23).

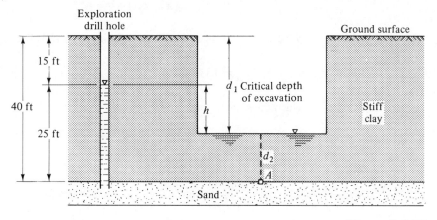

Exploration
drill hole

Ground surface

15 ft

d_1 Critical depth
of excavation

40 ft

h

Stiff
clay

25 ft

d_2

A

Sand

Figure P-5.23

The boiling and cracking of the bottom in the excavation will occur when the   *Solution*
intergranular pressure at the surface of the sand becomes equal to the upward
seepage pressure.

If d_1 is the excavation depth and d_2 is the distance from the bottom of hole to
the top of the sand layer in Fig. P-5.23, the respective pressures at the point A are

The intergranular pressure, $\bar{p} = (112 - 62.4)d_2$ (a)

The seepage pressure, $p_s = (25 - d_2)62.4$ (b)

The critical condition given by $\bar{p} = p_s$ is in full form

$$(112 - 62.4)d_2 = (25 - d_2)62.4$$

from which

$$d_2 = 14 \text{ ft}$$

Consequently, if

$$d_1 = 40 - 14 = 26 \text{ ft}$$

the boiling and/or cracking occurs and $d_1 = 26$ ft is the critical depth of excavation.

5.24   Using the relationships of Table P-1.23, compute the critical hydraulic gradient
i_c, (5-15), for the dense sand of porosity $n = 0.34$ and specific gravity $G_s = 2.7$.

From Table P-1.23, the unit weight of saturated soil in terms of specific gravity   *Solution*
of solid particles, porosity, and unit weight of water is

$$\gamma_{\text{sat}} = G_s - n(G_s - 1)\gamma_0 = 2.7 - (0.34)(2.7 - 1)(62.4) = 132 \text{ lb/ft}^3$$

The submerged unit weight is

$$\gamma_{\text{sub}} = \gamma_{\text{sat}} - \gamma_0 = 132 - 62.4 = 69.6 \text{ lb/ft}^3$$

Finally, the critical hydraulic gradient given by (5-15) is

$$i_c = \frac{\gamma_{\text{sub}}}{\gamma_0} = \frac{69.6}{62.4} = 1.12$$

at which the sand becomes unstable ($\gamma_0 \cong \gamma_w$).

5.25   Following the procedure of the preceding problem, compute the critical hydraulic gradient given by (5-15) for the loose sand of void ratio $e = 0.85$ and specific gravity $G_s = 2.7$.

Solution   As in Problem 5.24, the unit weight of saturated soil is

$$\gamma_{\text{sat}} = \frac{G_s + e}{1 + e} \gamma_0 = \frac{2.7 + 0.85}{1.0 + 0.85}(62.4) = 119.8 \text{ lb/ft}^3$$

The submerged unit weight is

$$\gamma_{\text{sub}} = \gamma_{\text{sat}} - \gamma_0 = 119.8 - 62.4 = 57.4 \text{ lb/ft}^3$$

Finally, again the critical hydraulic gradient, given as before by (5-15), is

$$i_c = \frac{\gamma_{\text{sub}}}{\gamma_0} = \frac{57.4}{62.4} = 0.92$$

5.26   Two soils (1) and (2) are placed above each other in a constant-head permeameter tube, as shown in Fig. P-5.26. The specific gravities and void ratios of these soils are $G_{s1} = 2.65, G_{s2} = 2.69, e_1 = 0.60, e_2 = 0.69$. If 25 percent of the hydraulic head is lost by the upward flow through soil 1, compute the critical hydraulic gradient at which the instability occurs.

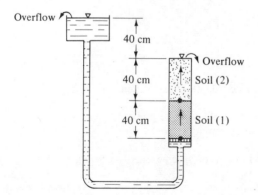

Figure P-5.26

The hydraulic gradients for these two soils are

$$i_1 = \frac{h_1}{L_1} = \frac{(0.25)(40)}{40} = 0.25$$

$$i_2 = \frac{h_2}{L_2} = \frac{(0.75)(40)}{40} = 0.75$$

The gradient in soil 2 is three times larger than the gradient of soil 1. Consequently, soil 2 instability must precede the soil 1 instability.

The critical hydraulic gradient for soil 2 is, by (5-15),

$$i_c = \frac{\gamma_{sub}}{\gamma_w} = \frac{\gamma_{sat} - \gamma_w}{\gamma_w}$$

and in terms of

$$\gamma_{sat} = \frac{G_s + e}{1 + e}$$

given in Table P-1.23,

$$i_c = \frac{(G_s + e)/(1 + e) - \gamma_w}{\gamma_w} = \frac{G_s + 1}{1 + e} = 1.0$$

where G_s and e for soil 2 are given values of this problem.

Hence the hydraulic head lost through soil 2 and causing the critical condition is

$$h_{c2} = i_{c2} L_2 = (1.0)(40) = 40 \text{ cm}$$

Since this head is equivalent to 75 percent of the total head loss through both samples, the total hydraulic head above surface of soil 2 that will produce a quick condition of instability is

$$h_c = \frac{h_{c2}(i_1 + i_2)}{i_2} = \frac{(40)(100)}{75} = 53.3 \text{ cm}$$

This condition is, of course, not possible in the system of Fig. P-5.26.

Field Permeability Tests and Theory of Wells

5.27 A pumping test was made in a pervious sandy soil extending to a depth of 50 ft, where an impermeable stratum was encountered (Fig. 5-8). The original ground water table was at the ground surface. Observation wells were installed at radial distances of 10 and 30 ft from the pumping well. A stable yield of 50 gal/min has been attained from the well after 24 hours. The drawdowns at the two observation wells were 5.0 and 1.0 ft. The porosity of the pervious sand material in which the pumping test is conducted is 0.30. Calculate the coefficient of permeability of the sand material.

Solution This is an ordinary perfect well (Fig. 5-8), where

$$Q = 50 \text{ gal/min} \quad \text{or} \quad 6.68 \text{ ft}^3/\text{min}$$

$$h_1 = (50 - 5.0) = 45 \text{ ft} \qquad h_2 = (50 - 1.0) = 49 \text{ ft}$$

$$r_1 = 10 \text{ ft} \qquad r_2 = 30 \text{ ft}$$

Substitution of the preceding quantities in (5-16) yields

$$k = \frac{(2.3)(6.68) \log (30/10)}{\pi[(49)^2 - (45)^2]} = 0.00621 \text{ ft/min}$$

5.28 For the system of Problem 5.27, compute the maximum radius of influence of the well.

Solution The radius of influence R is calculated by (5-17), where

$$t = 24 \text{ hr} = 1440 \text{ min}$$

$$n = 0.30 \qquad k = 0.00621 \text{ ft/min}$$

Thus

$$R = \sqrt{\frac{(12)(1440)}{(0.30)}} \sqrt{\frac{(6.68)(0.00621)}{\pi}} = 81.37 \text{ ft}$$

5.29 A pumping test was made in a pervious strata extending to a depth of 40 ft, where an impervious stratum was encountered. The original water table was at the ground surface. The test well has a diameter of 2.0 ft. A yield of 120 gal/min was established by a steady pumping that produced a maximum drawdown of 25 ft in the test well. Assuming the radial distance of 500 ft to where the drawdown is zero, calculate the coefficient of permeability k.

Solution Since the radius of influence R is assumed in this problem, equation (5-16) can be modified as

$$k = \frac{2.3Q \log (R/r_0)}{\pi(H^2 - h_0^2)} \tag{a}$$

Then with

$$Q = 120 \text{ gal/min} = 16.04 \text{ ft}^3/\text{min}$$

$$R = 500 \text{ ft} \qquad r_0 = 1 \text{ ft}$$

$$H = 40 \text{ ft} \qquad h_0 = 40 - 25 = 15 \text{ ft}$$

the in situ permeability is

$$k = \frac{(2.3)(16.04)(\log 500/1.0)}{\pi[(40)^2 - (15)^2]} = 0.0231 \text{ ft/min} \tag{b}$$

5.30 Assuming that $R = 1000$ ft in Problem 5.29, determine the coefficient of permeability k and compare the result of this problem with its equivalent in the preceding problem.

Using (a) of Problem 5.29 in terms of parameters used in that problem but with *Solution* $R = 1000$ ft, the permeability coefficient becomes by (5-16)

$$k = \frac{(2.3)(16.04)(\log_{10} 1000/1.0)}{\pi[(40)^2 - (15)^2]} = 0.0256 \text{ ft/min}$$

which is only about 11 percent greater than the value obtained for $R = 500$ ft.

Since the radius of influence R is usully very large when compared to the radius of the test well r_0, the value of $\log (R/r_0)$ will vary only over a narrow range with the variation of R.

Therefore, in some field problems, an approximate value of k may be determined using this general relationship (5-16) and assuming a reasonable value of R. It will not be necessary in the latter case to construct observation wells. This approximate method proved to be satisfactory in many cases.

5.31 A pumping test using a perfect normal ordinary well was made in a sandy material that is 25 ft thick and has a coefficient of permeability of 1 ft/min. The radius of the well is 0.5 ft and the maximum drawdown is 10 ft. For the radius of influence $R \cong 300$, determine the rate of discharge.

By (5-18) and with *Solution*

$$k = 1.0 \text{ ft/min}, \quad H = 25 \text{ ft}, \quad h_0 = 25 - 10 = 15 \text{ ft}$$
$$R \cong 300 \text{ ft}, \quad r_0 = 0.5 \text{ ft}$$

the rate of discharge

$$Q = \frac{(\pi)(1.0)[(25)^2 - (15)^2]}{2.3 \log_{10} (300/0.5)} = 196.6 \text{ ft}^3/\text{min}$$

5.32 For the system of Problem 5.31, sketch the depression line with the origin of the ground water table at the ground surface.

The depression curve is given by (5-19). Its graphical representation is shown in *Solution* Fig. P-5.32 and its numerical values are recorded in Table P-5.32. The depression curve in Fig. P-5.32 is determined by (5-19), by assuming different values of x and calculating the corresponding value of elevation y.

5.33 The maximum drawdown produced by a steady pumping discharge from a perfect artesian well is 10 ft. The thickness of aquifer is 30 ft; coefficient of permeability of aquifer pervious soil is 1.0 ft/min, and the radius of the well is 0.5 ft. Maximum influence of the well is 700 ft. Compute the rate of discharge from the well Q_1.

TABLE P-5.32

Radial Distance from Center of Test Well x (ft)	Elevation above Bottom or Base of Well y (ft)	Drawdown $d = H - y$ (ft)
$R = 300$	$H = 25$	0
250	24.8	0.2
200	24.5	0.5
150	24.1	0.9
100	23.6	1.4
50	22.7	2.3
10	20.3	4.7
5	19.2	5.8
$r_0 = 0.5$	$h_0 = 15.0$	$d_{max} = 10.0$

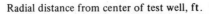
Radial distance from center of test well, ft.

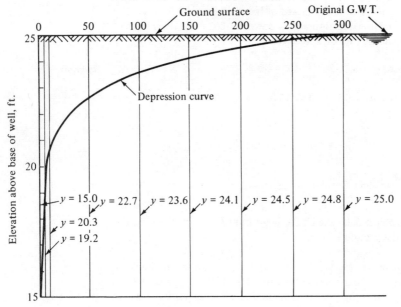

Figure P-5.32

Solution By (5-22) and in terms of

$$k = 1.0 \text{ ft/min}, \quad S = 30 \text{ ft}, \quad d_{max} = 10 \text{ ft}, \quad R = 700 \text{ ft}, \quad r_0 = 0.5 \text{ ft}$$

the rate of discharge is

$$Q_1 = 2.73 \frac{(1.0)(30)(10)}{\log(700/0.5)} = 260.4 \text{ ft}^3/\text{min}$$

5.34 Calculate the rate of flow into the artesian well of the preceding problem for the maximum drawdown of 20 ft. Compare the discharge Q_1 of Problem 5.33 with the result Q_2 of this problem.

By (5-22) with $d_{\max} = 20$ ft, the rate of discharge is   *Solution*

$$Q = 2.73 \frac{(1.0)(30)(20)}{\log(700/0.5)} = 520.7 \text{ ft}^3/\text{min}$$

This result indicates that increasing the drawdown by 100 percent produces an increase of 100 percent in the discharge, or that various discharges from the same well are directly proportional to their maximum drawdowns, or that

$$\frac{Q_1}{Q_2} = \frac{d_{1\,\max}}{d_{2\,\max}}$$

SUPPLEMENTARY PROBLEMS

Permeability

5.35 A soil sample has a permeability coefficient $k = (1.10)10^{-2}$ cm/sec. The water temperature during the test was found to be 15°C. What would be the coefficient of permeability of this soil at a standard temperature of 20°C?

k at 20°C $= 1.23 \times 10^{-2}$ cm/sec   *Answer*

5.36 In a field pumping test, the coefficient of permeability of a granular deposit was found to be $k = 1.28$ ft/min. What would be the value for the effective grain size of this granular material?

$D_{10} \cong 0.809$ mm   *Answer*

5.37 A sample of soil was tested in a constant-head permeameter. Its length and diameter were 12 and 5 cm respectively. If the head was maintained at 24 cm and the amount of water collected was 160 cm³ in 2 minutes, compute the coefficient of permeability of this soil.

$k = 0.034$ cm/sec   *Answer*

5.38 A sample of fine sand 50 cm long and 10 cm wide was tested in a constant-head permeameter. A constant head was maintained, and the amount of overflow collected was at the rate of 200 cm³/min. The loss of head between two points 25 cm apart was 40 cm. Compute the coefficient of permeability.

$k = 0.0266$ cm/sec.   *Answer*

5.39 If the same sample in Problem 5.27 was used in a falling-head permeameter that has a diameter of 2.0 cm for the standpipe, compute the time during which the level of water in the standpipe falls from 100 to 50 cm. Comment on the results.

Answer $t = 52.2$ seconds, which is very fast and not practical. This result illustrates the fact that the falling-head permeameter is more suitable to perform tests on relatively low permeability soils.

5.40 A falling-head permeameter was used to determine the permeability of a sandy-clay soil sample having a cross-sectional area of 12.56 cm² and a length of 18 cm. The cross-sectional area of the permeameter standpipe is 2 cm². The time during which the level of water in the standpipe dropped from 100 to 50 cm was 4 minutes. Determine the coefficient of permeability of this material.

Answer $k = 0.496$ cm/min

5.41 A soil sample 12 in. long and 2 in. in diameter was placed in a falling-head permeameter that has a standpipe of 0.5 in². The water during the test was falling from 50 to 25.5 in. in a period of 14 hours, 53 minutes, and 4 seconds. Calculate the coefficient of permeability of the soil.

Answer $k = 1.439 \times 10^{-3}$ in./min

Seepage

5.42 A sheet pile similar to the one in Fig. P-5.42 is driven into a permeable soil to a depth of 20 ft. Water stands at a depth of 20 ft on one side of the sheet pile and is continuously drained from the other side, where it stays dry all the time. Sketch the flow net on the downstream side of the pile and show how it can be used to determine whether a prism 10 ft wide, 20 ft deep, and 1 ft thick on the downstream side of the sheet pile will be stable under the seepage force. Assume a saturated unit weight of 118 lb/ft³ for the soil.

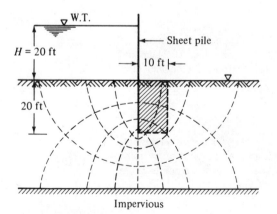

W.T.

Sheet pile

$H = 20$ ft

10 ft

20 ft

Impervious

Figure P-5.42

Total number of equipotential drops (from flow net Fig. P-5.42) is 8. Mean loss of *Answer* head through prism $= (\frac{3}{8})H = (\frac{3}{8})20 = 7.5$ ft. Seepage force/unit volume of soil in prism $= i\gamma_0 = (7.5/20)62.4 = 23.50$ lb/ft^3. Volume of soil susceptible to failure in prism $= (20)(10)(1) = 200$ ft^3. Total upward seepage force $= (200)(23.50) = 4700$ lb. Unit weight of submerged soil in prism $= (118 - 62.4) = 55.6$ lb/ft.^{3} Total weight of prism acting downward $= (55.6)(200) = 11120$ lb. Factor of safety against failure of the prism under the influence of the seepage force $= (11120/4700) \cong 2.15$.

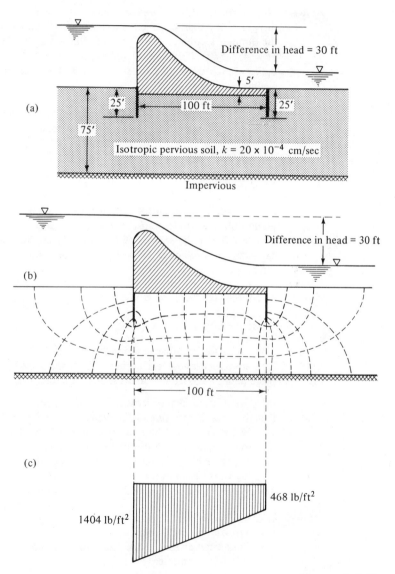

Figure P-5.43

5.43 Draw the flow net for seepage through the pervious foundation of the concrete dam shown in Fig. P-5.43a to scale 1 in. = 30 ft.

Answer Fig. P-5.43b

5.44 For the dam of Problem 5.42, plot by means of the pizometric surface the distribution of hydrostatic uplift pressure along the base of the dam. Assuming a reasonable value for the submerged unit weight of the concrete material, state whether, in your opinion, the dam is safe against uplift.

Answer Hydrostatic uplift pressure distribution is shown in Fig. P-5.43. Assuming a submerged unit weight of concrete to be $\cong (150 - 62.4) = 87.6 \, \text{lb/ft}^3$, and considering the thinnest section of dam with 5-ft thickness, the downward gravity load would be $= 438 \, \text{lb/ft}^2$. The upward uplift pressure $= 700 \, \text{lb/ft}^2$. Factor of safety against uplift $= 438/700$, which is less than 1.0. Thus this toe section of dam is not safe against uplift pressure.

5.45 If the length of the dam in Problem 5.43 is 350 ft, compute the seepage loss through the foundation in cubic feet per second.

Answer Seepage loss under entire length of dam Q is 0.23 ft^3/sec or 827 ft^3/hr

Stratified Soils

5.46 Consider a system of soil composed of three layers that have coefficients of permeability as follows: $k_1 = 0.06, k_2 = 0.075$, and $k_3 = 0.135$ (all in feet per minute). If the thicknesses of the three layers are 16, 10, and 14 ft, respectively, calculate the coefficient of permeability of the compound structure in the lateral direction.

Answer k_x for the compound structure $= 0.09 \, \text{ft/min}$

Pressure Effects of Moisture Flow

5.47 In a test arrangement similar to that shown in Fig. P-5.26, two soils (1) and (2) were placed above each other in a constant-head permeameter tube. Flow takes place under a constant head, as shown in figure. Answer the following questions:

(a) What is the total or hydraulic head at point A?
(b) What is the pressure of the pizometric head at point A?
(c) If 30 percent of the hydraulic head is lost when water flows upward through soil sample 1, what are the total head and the pizometric head at point B?
(d) If permeability of soil 1 is 0.045 cm/sec, what quantity of water would be flowing through unit area of soil per second?
(e) What is the coefficient of permeability of soil 2?

Answer (a) Hydraulic head at $A = 40$ cm
      (b) Pizometric head at $A = 120$ cm
      (c) Total head at $B = 28$ cm and pizometric head at $B = 68$ cm
      (d) $q = 0.0135$ cm^3/sec
      (e) k for soil 2 $= 0.0193$ cm/sc

5.48 If the hydraulic pressure is increased in the situation illustrated by Problem 5.47, determine the hydraulic head at which instablility or quick condition occurs. The specific gravities of soils 1 and 2 are 2.65 and 2.70 respectively; the void ratios of soils 1 and 2 are 0.55 and 0.70 respectively.

Critical hydraulic head is 57 cm   *Answer*

Field Permeability Tests and Theory of Wells

5.49 Compute the rate of discharge Q_1 into an ordinary perfect well that was drilled in a pervious sandy soil that has a coefficient of permeability of 0.90 ft/min. The diameter of the well is 0.6 ft and the maximum drawdown in the well is 10 ft. The original ground water table has an elevation of 21 ft above the bottom of well, and the radius of influence was estimated to be about 2400 ft.

Rate of discharge $Q_1 = 6048$ ft^3/hr   *Answer*

5.50 In Problem 5.49, compute the rate of discharge into the well if its diameter was increased to 1.2 ft. Compare the discharge values of Q_1 in Problem 5.49 and of Q_2 in this problem.

Rate of discharge $Q_2 = 6546$ ft^3/hr; $Q_2/Q_1 = 1.08$, or that increasing the radius of *Answer* the well by 100 percent produced an increase in the rate of discharge by only about 8.2 percent.

volume
change

6

6.1 INTRODUCTION

Concept

In general, the deformation of a three-phase matter, such as soil, is defined as its *volume change*. If the enumeration of causes is restricted to the domain of physical processes, the main causes of deformation are the *loads*, the *change in temperature*, and the *change in moisture content*.

The occurrence of deformation in soils is conditioned by the existence of a bond between particles and/or the confinement of loose particles by lateral boundary constraints. If the bond is destroyed and/or if the constraints are removed, the soil (under the action of loads) will separate (break up) or scatter. Some soils (such as the skeletal soils) that are able to absorb the effect of large static loads collapse partially or completely under the action of dynamic loads (shakedown).

Definitions

Several frequently encountered terms related to the process of volume change are defined below.

1. The *compressibility* (volume change flexibility) is the volume decrease due to a unit load.

2. The *consolidation* (volume change velocity) is the rate of decrease in volume with respect to time.

3. The *temperature expansion* or *contraction* is the change in volume of soil due to a change in temperature.

4. The *shrinkage* is the volume contraction of soil due to reduction in water content.

5. The *swelling* is the volume expansion of soil due to increase in water content.

Foundation Settlement

Since the process of consolidation of loaded soils is invariably accompanied by the variation in temperature and moisture content, the volume changes defined by terms (2) to (5) occur jointly, in succession and/or in an alternating pattern. As such, they are the prime cause of foundation displacement (settlement, rotation, uplift) and of the resulting distortions of the structure above the ground. The study of these parameters of displacement, pertinent to the analysis and design of foundations, is presented in this chapter.

6.2 COMPRESSIBILITY

Mechanism of Deformation

If a soil sample is in a state of stress, the resulting volume change may be attributed to some or all of the following factors :

1. The relocation of solids (their rearrangement).
2. The deformation of solids (their change in shape).
3. The deformation of pore water and air.
4. The extrusion of pore water and air.

In cohesionless soils and to a limited extent in clays, the major factor of volume change is the distortion of grains, which is largely elastic and consequently reversible. In turn, the rearrangement of grains and their fracture at points of contact contribute to volume changes that are irreversible.

In cohesive soils the major factors of volume change are the slipping of particles into a new, denser position, the bending of sheetlike particles, and the extrusion of pore water (to lesser extent, of pore air). The clay soil exhibits the typical rheological characteristics (slow time-dependent deformations) described in Section 6.3.

Immediate Elastic Displacement

The immediate elastic displacement of the surface of the soil (which occurs without change in water content) caused by the surface load P distributed over a rectangular area of length l and width b in the form of $p = P/lb$ is given by the *Terzaghi formula* as

$$\Delta_i = pb\frac{1 - \mu^2}{E_0}\phi \qquad (6\text{-}1)$$

where Δ_i = the vertical settlement

p = the intensity of the uniformly distributed load

μ = Poisson's ratio of soil

E_0 = the elastic modulus of subgrade

ϕ = influence value, depending on the shape or size of loaded area

$$= \frac{1}{\pi}\left[m \log \left(\frac{1 + \sqrt{m^2 + 1}}{m} \right) + \log \left(m + \sqrt{m^2 + 1} \right) \right]$$

$$m = \frac{l}{b}$$

The modulus of subgrade E_0 is an experimental constant, the value of which may be estimated from Table 6–1 (Problem 6. 1).

Although several other formulas have been developed for this purpose, it is quite obvious that the continuum mechanics approach cannot supply the final correct answer to the volume change governed by factors enumerated at the onset of this section.

TABLE 6-1   *Values of Modulus of Subgrade*

Modulus E_0 in pounds per square inch

Subgrade	Solid Base	Well Graded	Poorly Graded
Sandstone	1,400,000–1,800,000	–	–
Limestone	600,000–800,000	–	–
Lime, air-slaked	–	190,000	150,000
Sand and gravel	–	40,000–50,000	30,000–40,000
Medium sand	–	20,000–30,000	14,000–18,000
Clay	–	12,000–15,000	11,000–12,000

Compressibility Test

The first rational clarification of the soil compression behavior has been devised by Terzaghi (1925) and is known as the Terzaghi compressibility test. The apparatus used for this purpose, the testing procedure, and a set of typical results from this test are given in Problem 6.2. The most distinct characteristic of Terzaghi's test is the incremental loading pattern. Loads are applied in a series of $\frac{1}{4}, \frac{1}{2}$, 1, 2, 4, 8, 16kg/cm². In making the test, the first load of $\frac{1}{4}$kg/cm² is sustained until the deformation flow has practically stopped. Then the void ratio of the sample is determined, the load is doubled, and the process is repeated.

Test Specimens

Two types of specimens are used in this test, the undisturbed specimen and the remolded specimen.

1. The *undisturbed specimen* is a sample cut from the natural formation in the size, exactly fitting the testing devise. This sample has virtually all properties of the natural formation except the loading condition (the confining pressure of the surrounding medium is partially or completely removed).
2. The *remolded specimen* of a reworked sample of soil with pretested manipulations (change in structure, saturation, etc.).

The merits and drawbacks of using these specimens are discussed in Problems 6.3 and 6.4.

Compressibility Diagram

Plotting the results of the compressibility test (pressure versus void ratio) leads to the diagram of Fig. 6–1a, called the virgin curve. The curve is concave,

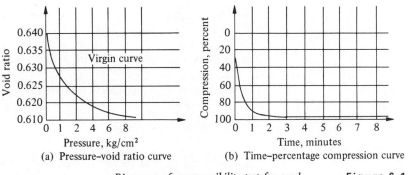

(a) Pressure–void ratio curve

(b) Time–percentage compression curve

Diagrams of compressibility test for sand **Figure 6-1**

revealing an anomalic phenomenon : the rate of compression is decreasing as the load increases. The time compression in percent curve of Fig. 6–1b shows that the major part of compression occurs almost instantaneously. If a similar curve is plotted for clay, the compression does not occur instantaneously and a time lag exists between the cause and the effect. Problem 6.5 describes the compressibility of undisturbed and remolded clay samples and the hydrostatic and plastic lag of their delayed deformation (see also Section. 6.4).

Decompression and Recompression

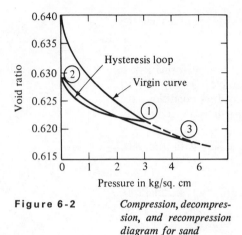

Figure 6-2   *Compression, decompression, and recompression diagram for sand*

In addition to the question of the time rate at which the compression takes place, the volume change caused by unloading and reloading is of practical significance. If the pressure is increased to a certain magnitude and then completely released, the soil will not return to its original volume but will expand along the *decompression curve* (Fig. 6-2, curve 12). If the same pressure is applied again, the soil will contract along the *recompression curve* (Fig. 6-2, curve 23) and practically join the virgin curve. The area between the decompression and recompression curve is called the *hysteresis loop*. It differs in slope and size for different soils but has the same general shape.

Properties of Diagrams

Three terms related to certain properties of the compressibility diagram are defined below.

1. The coefficient of compressibility is the slope of the virgin curve, given as

$$a_v = -\frac{de}{dp} \cong -\frac{\Delta e}{\Delta p} = -\frac{e_0 - e}{p_0 - p} \tag{6-2}$$

where e_0 is the void ratio at pressure p_0 and e is the void ratio at pressure p ($e < e_0$, $p > p_0$).

2. The compression index appears as a constant C_c in the relationship

$$e = e_0 - C_c \log \frac{p}{p_0} \tag{6-3}$$

which is the equation of the straight portion of the virgin curve plotted with the logarithm of pressure as the abscissa and the void ratio as the ordinate.

3. The approximate compression index for remolded clay is given by the *Skemton formula* as

$$C_c' = 0.007 \, (LL - 10\%) \tag{6-4}$$

$$e - e_0 = -c \log \frac{p}{p_0}$$

$$e / \frac{e - e_0}{c} = \log \frac{p}{p_0} \qquad e_0 - e$$

where LL is the liquid limit expressed as a percentage. This approxima-tion is within the error of $\pm 30\%$ compared to C_c in (6-3). For an ordi-nary clay of medium or low sensitivity, the value of C_c corresponding to the field consolidation line is approximately equal to $1.3C_c'$.

$$C_c = 1.3C_c' = 0.009(\text{LL} - 10\%) \tag{6-5}$$

4. The swelling index appears as a constant C_s in the relationship

$$e = e_1 + C_s \log \frac{p}{p_0} \tag{6-6}$$

which is the equation of the decompression curve plotted semilogarithmi-cally as in (2). The new term e_1 is the void ratio at pressure release.

The application of (6-2) to (6-6), plus additional discussion, is presented in Problems 6.6, 6.7, and 6.8.

Correlation

The compressibility of a soil layer in the field can be determined with reason-ably good accuracy by proportion from a sample of the same soil.

For the soil layer of a known void ratio e and thickness H, the change in thickness ΔH corresponding to the change of void ratio Δe is

$$\Delta H = \frac{H \, \Delta e}{1 + e_0} \tag{6-7}$$

where Δe is obtained by testing the sample (Problem 6.9).

Frequently, however, it may be desirable to approach the problem in terms of pressure changes and compressibility rather than in terms of void ratio changes. Problem 6.6 shows this alternate approach.

6.3 CONSOLIDATION

Process

The study of the compressibility of clay in Section 6.2 revealed a slow rate of compression, attributed to the gradual adjustment of the position of grains, but primarily to the slow dissipation of pore water related to the very low permeability of the material. The rate of the volume change (in this process) is called the *consolidation*, and the history of the process is usually given by the time-compression diagram of Fig. 6-3, which is characterized by three compression zones (initial compression zone, primary consolidation zone, sec-ondary consolidation zone).

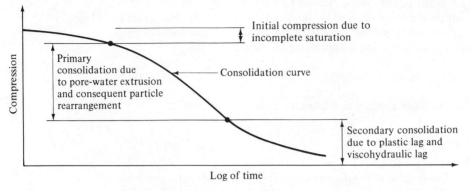

Initial compression due to
incomplete saturation

Primary
consolidation due
to pore-water extrusion
and consequent particle
rearrangement

Consolidation curve

Secondary consolidation
due to plastic lag and
viscohydraulic lag

Log of time

Figure 6-3   *Time-compression diagram for clay*

Consolidation Theory

The first scientific theory of soil consolidation (presented in an analytical form) is credited to Terzaghi (1925) and is frequently designated by his name. This theory serves as the basis for the determination of thickness change in a given layer of soil at a given time under the prescribed pressure.

The assumptions of the theory are

1. The soil considered is a homogeneous, fully saturated clay-water system.
2. The compressibility of soil grains and water is small and can be disregarded.
3. The compression is one dimensional and in the direction of water flow, governed by Darcy's law (1) (laminar flow).
4. The cause-effect relationship is linear, allowing the application of the small-strain theory.

Governing Equation

From the equality of the rate of change in volume due to water outflow and in volume due to changes in pore volume, the governing equation of the consolidation theory is

$$c_v \frac{\partial^2 u}{\partial z^2} = \frac{\partial u}{\partial t} \tag{6-8}$$

where u = hydrostatic excess (see below),

z = position coordinate,

t = time,

c_v = coefficient of consolidation (see below) (Problem 6.10).

The hydrostatic excess u is

$$u = \gamma_w h \tag{6-9}$$

where h is the hydraulic head with respect to the ground water table above the consolidating layer.

The coefficient of consolidation c_v is

$$c_v = \frac{k(1 + e_0)}{a_v \gamma_w} \qquad (6\text{-}10)$$

where k = permeability,

$\quad\quad e_0$ = void ratio,

$\quad\quad \gamma_w$ = unit weight of water,

$\quad\quad a_v$ = coefficient of compressibility given by (6-2).

General Solution

The solution of (6-8) involves finding the $u = \psi(z, t)$ that satisfies both the differential equation and the boundary conditions associated with the problem. This can be done by any of the methods available for the solution of this partial differential equation. A set of such solutions is given in Problems 6.11 and 6.12.

Percentage of Consolidation

The consolidation ratio is

$$U_z = 1 - \frac{u}{u_i} \qquad (6\text{-}11)$$

where u is the hydrostatic excess at $t = t$ and u_i is the hydrostatic excess at $t = 0$.

The formulation of (6-11) in terms of the integral of (6-8) over the depth of consolidation layer gives the average values of U_z, denoted as

$$U = 1 - \frac{\int_0^{2H} u \, dz}{\int_0^{2H} u_i dz} = f(T_v) \qquad (6\text{-}12)$$

where $2H$ is the depth of the layer and T_v is the time factor, given as

$$T_v = \frac{c_v}{H^2} t \qquad (6\text{-}13)$$

where t is the time and c_v is the coefficient of consolidation given by (6-10).

The functional equation (6-12) for every open layer of thickness $2H$ is given by the curve C_1 of Fig. 6-4. If the base of the layer is impermeable and the consolidation pressure is zero at the bottom, the functional equation (6-12) is given by the curve C_2. Finally, if the base of the layer is impermeable and

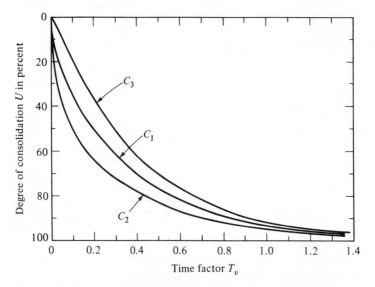

Figure 6-4   *Relation between the time factor* T_v *and the degree of consolidation* U.

K. Terzaghi and O. K. Frohlich, *Theorie der Setzung von Tonschichten*, Deuticke, Leipzig, 1936, p. 136.

the consolidation pressure is zero at the top, the functional equation (6-12) is given by the curve C_3. The application of these relationships in relation to specific boundaries is given in Table 6-2.

TABLE 6-2

DEGREE OF CONSOLI-DATION	TIME FACTOR T_v		
U %	Condition 1	Condition 2	Condition 3
10	0.008	0.003	0.047
20	0.031	0.009	0.100
30	0.071	0.024	0.158
40	0.126	0.048	0.221
50	0.197	0.092	0.294
60	0.287	0.160	0.383
70	0.403	0.271	0.500
80	0.567	0.440	0.665
90	0.848	0.720	0.940
100	∞		

NOTE: For two-way drainage—Condition 1, length of drainage path is taken as 1/2 the thickness of the layer. *Compiled from K. Terzaghi and O. K. Frohlich*, Theorie der Setzung von Tonschichten, *Deuticke, Leipzig, 1936, p. 166.*

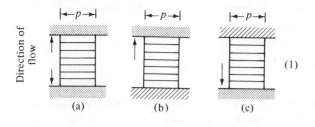

(1)

Boundary Condition 1: (*a*) *Double drainage and all linear distributions of consolidating pressure*; (*b*) *and* (*c*) *Single drainage and uniform distribution of consolidating pressure.*

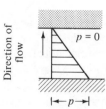

(2)

Boundary Condition 2: *Drainage from only one side and the pressure distribution is as shown.*

(3)

Boundary Condition 3: *Drainage from only one side and the pressure distribution is as shown.*

6.4 RHEOLOGY OF CONSOLIDATION

Secondary Consolidation

The theory of consolidation (advanced by Terzaghi some four decades ago) presents a crude yet very useful analytical tool. It makes a good interpretation of the primary consolidation process within the limits of its assumptions. The secondary consolidation, as shown by the third zone of the consolidation curve in Fig. 6-3, is not yet well understood and its effect is not included in Terzaghi's theory. Yet it can be a major part of the compression in highly compressive clays and cannot be disregarded.

Linear Rheological Models

The disagreement between Terzaghi's theory and the behavior of a test sample was first analytically investigated by Taylor and Merchant (1939) and led to the introduction of the first linear rheological model (Burger model, Fig. 3-9) in soil mechanics. Since then other rheological models have been developed and many more are in the process of development.

Their great advantage lies in the possibility of using the equations of elasticity with Lame's coefficients replaced by the parameters of the soil. In these models, the elastic response of the soil is represented by a linear spring, and the water viscosity of soil is approximated by the dashpot, as shown in the works of Tan (1957), Gibson and Lo (1961), Shiffman (1961), Anagnosti (1962), and others (Problems 6.16, 6.17).

Although these parameters (k = constant of proportionality, $1b/in.^2$, and η = constant of viscosity, $1b/in.^2 \times t$) may be easily specified from experiments for a given physical situation in soil, in general, for a different situation (load, rate of loading, moisture rate of pore-water dissipation) the *same constants do not apply*. In order for a model to be reliable, it must conform or reasonably approximate the real soil system. Since soil is a nonlinear material, it cannot be represented to a good degree of accuracy by a linear model.

Nonlinear Rheological Models

Consequently, we are led to the conclusion that k and η must be treated as functions of causes, properties of soil and time, and, as such, inserted in the respective differential equation. This refinement brings the mathematical model to a closer agreement with the physical model, but the resulting partial differential equation becomes nonlinear and its solution is a formidable task. The application of numerical methods in conjunction with computer programming appears to be the most feasible approach to this problem. For additional information on the rheological aspects of soil mechanics and recent advances in this area, the reader is referred to the work of Šuklje (1969).

SOLVED PROBLEMS

Immediate Elastic Displacement

6.1   Calculate the immediate elastic displacement of the surface of a soil mass that is classified as *SP* according to the Unified Soil Classification System (see Chapter 2). The soil is loaded with a rectangular slab 20 ft long and 10 ft wide, and loaded with a uniform pressure of 40 $1b/in.^2$

By (6-1), the immediate elastic displacement is   *Solution*

$$\Delta_i = pb \frac{1 - \mu^2}{E_0} \phi \tag{a}$$

To calculate the value of ϕ in equation (a), the shape factor m must be determined first.

$$m = \frac{l}{b} = \frac{20}{10} = 2.0 \tag{b}$$

Then the influence factor is determined.

$$\phi = \frac{1}{\pi}\left[m \log \frac{1 + \sqrt{m^2 + 1}}{m} + \log (m + \sqrt{m^2 + 1}) \right]$$

$$= \frac{1}{\pi}\left[2 \log \frac{1 + \sqrt{4 + 1}}{2} + \log (2 + \sqrt{4 + 1}) \right] = 0.33 \tag{c}$$

From Table 6-1, assume that a value of E_0 for an SP soil is 14,000 lb/in.² Assuming that there will be no volume change in the soil, the value of Poisson's ratio μ may be assumed to be $= 0.5$. Values of p and b are given in the problem; thus substituting in equation (6-1) yields the value of the immediate elastic deformation

$$\Delta_i = (40)(10)(12) \left[\frac{1 - (0.5)^2}{14,000} \right](0.33) = 0.848 \text{ in.}$$

Compressibility Test

6.2   Describe the consolidation test apparatus and outline the test procedure. Also, show some typical results from this test.

The apparatus used in the laboratory test is shown diagramatically in Fig. P-6.2a   *Answer* and b. The laboratory test is one dimensional in that, with a metal ring confining the sample, no lateral soil or water movement takes place. The size of specimens tested by different laboratories varies considerably between 0.75 and 1.50 in. in thickness and 1.80 to $4\frac{7}{16}$ in. in diameter. For routine measurements, diameters of 2.50 in. with an initial thickness of 1 in. are used. The apparatus may be of the fixed-ring, Fig. P-6.2a, or the floating-ring type, Fig. P-6.2b. The advantage of the fixed-ring device is that it can be used to measure the coefficient of permeabililty of the sample as it is being tested. The advantage of the floating-ring device is that it reduces the friction loss along the sides of the sample between the soil and the ring.

Two sets of information are usually produced from the consolidation test:

1.   Void ratio-effective pressure relationship.
2.   Degree of consolidation-time relationship.

Test procedure is as follows:

1.   A load Δp_1 is applied to the sample and the change in thickness (compression) of the sample is read at suitable intervals, generally after $\frac{1}{2}$, 1,2, 4, 6, 8, 16, . . .

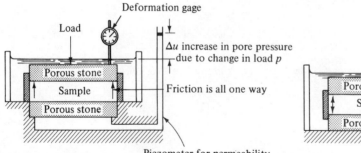

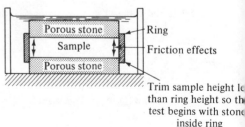

(a) Fixed-ring consolidometer

(b) Floating-ring consolidometer

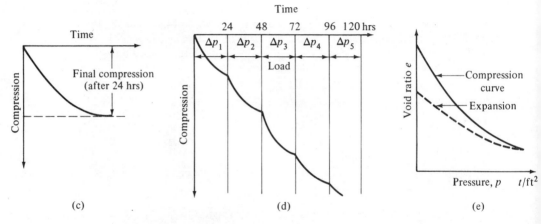

(c)

(d)

(e)

Figure P-6.2 *J. E. Bowles,* Engineering of Soils and Their Measurement, *McGraw-Hill, New York, 1970, p. 107.*

etc., minutes until enough time has elapsed to allow the settlement to be complete (usually 24 hours).

2. A graph of time compression is plotted as shown in Fig. P-6.2c.

3. Next, a load Δp_1 is increased to Δp_2 and plotting continues for a further 24-hour period. Another load increment is then added and plotting continues. The successive loads are usually $\frac{1}{4}, \frac{1}{2}, 1, 2, 4, 8, 16,$ and 32 t/ft². The limit of the final increment depends on the actual load that the soil would be subjected to under the foundation. The final graph will be as shown in Fig. P-6.2d.

4. If one is interested in knowing the swelling properties of the sample when the load is removed, the loads are removed from the specimen (after the loading cycle has been completed) in increments; 24-hour intervals elapse before each increment is removed. The dial readings are taken as in the loading process, until the load has been reduced to zero.

5. A complete void ratio-effective pressure for both loading and unloading conditions can then be plotted. The results of the swelling test are of practical use in the determination of upward movement of soil when the overburden load is removed as in the excavation for a cut or deep foundation. An example of the e/p compression curve is shown in Fig. P-6.2e.

6. The coefficient of permeability can be determined during the consolidation test by attaching a standpipe to the fixed-ring consolidometer (Fig. P-6.2a) and performing the falling-head permeability test when the consolidation has been completed under a particular increment. The permeability can also be determined indirectly by using (6-9).

6.3 Define the following classes of soil : (a) underconsolidated, (b) normally consolidated, (c) preconsolidated. Also, explain the effect of sample disturbance on the results produced in a laboratory consolidation test and show how this would affect the extrapolation on field consolidation from laboratory test data.

Definitions :   *Answer*

a. Underconsolidated soil is one that has not yet fully consolidated under the existing overburden pressure.
b. Normally consolidated soil is one that has never been subjected to an effective pressure greater than the existing overburden.
c. Preconsolidated soil is one that has been subjected to an effective pressure greater than the present overburden pressure.

Preconsolidation of clay stratum may occur for several reasons—for example, the existence during a period of time in the past of a heavier overburden that has been eroded or excavated.

A method of estimating preconsolidation pressure p_0' was suggested by Casagrande and is illustrated in Fig. P-6.3a. The figure shows the $e/\log p$ curve for an undisturbed clay sample. The point c of maximum curvature (or minimum radius) is determined; then a horizontal line is drawn through it. The bisector of the angle α

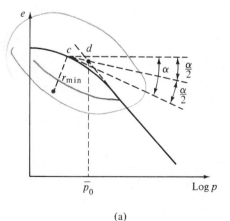

(a)

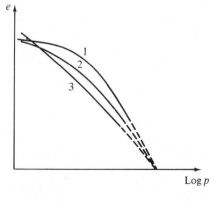

(b)

Figure P-6.3

between this line and the tangent to the curve intersects the upward continuation of the straight lower part at point *d*. The abscissa of *d* is assumed to be approximately the value of p_0'.

The effects of sample disturbance on the laboratory consolidation test are as follows:

1. For any given vertical pressure, the void ratio of a disturbed sample is lower than that of an undisturbed sample.
2. Sample disturbance makes it difficult to determine the preconsolidating load, using the method previously outlined and presented in Fig. P-6.3a.
3. The straight-line segment of the remolded compression curve is displaced downward from the undisturbed laboratory virgin curve.

The greater the disturbance of the sample, the more these effects will be pronounced in the laboratory test results. Figure P-6.3b illustrates this situation, where curve 1 is the $e/\log p$ curve for an undisturbed sample, curve 2 represents a slightly disturbed sample, and curve 3 represents a completely remolded or highly disturbed sample. Thus it is clear that the sample disturbance tends to destroy the evidence of preloading. Therefore if it is suspected that a clay deposit has been preloaded, laboratory samples for a consolidation test should be obtained with great care and the best possible sampling techniques should be used so that the sample is not disturbed.

6.4 Using a deformation-log time curve plotted from a laboratory consolidation test, show how the deformations corresponding to 0 and 100 percent consolidation can be determined.

Answer A method suggested by Casagrande is used to solve this problem. The steps are as follows:

1. Plot the deformation (dial reading) against the logarithm of the time for one of the load increments in the consolidation test, as shown in Fig. P-6.4.
2. The two straight portions of the curve are extended to intersect at 100 % *U*.

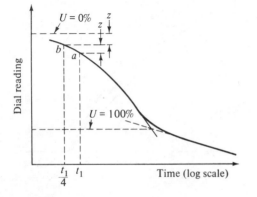

Figure P-6.4

3. The corrected value of 0 % U is obtained by assuming that the curved portion at the beginning of the graph is a parabola. A time t_1 is arbitrarily selected on the curve (the point should be at a location on the curve where less than 50 percent of the consolidation for the entire load increment has taken place). A second point is then selected corresponding to time $t_1/4$. The vertical distance on the curve between a and b (Fig. P-6.4) is z. A horizontal line is then drawn at a height equal to z above b. The ordinate of this horizontal line is the corrected zero reading corresponding to 0 % U.

Compressibility Diagrams

6.5 Sketch a typical pressure–void ratio, and a typical time compression in percent curves for a sample of saturated clay in a compressibility test. Compare these curves with the similar ones for sand in Fig. 6.1 and comment on the results of your comparison.

The required relationships are sketched in Fig. P-6.5a and b. Comparing them with the similar relationships for sand in Fig. 6-1a and b, the following comments can be made: *Answer*

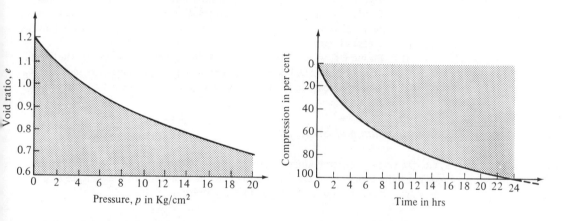

(a) Pressure-void ratio curve

(b) Time-percentage compression curve

Figure P-6.5

1. Curves representing pressure-void ratio for both sand and clay are similar in that each shows a decrease of the void ratio with the increase of pressure. However, the decrease for sand, especially if it is loose, will be more rapid at low values of pressure; then, as the pressure is built up, the slope of the curve will decrease.

2. Figure 6-1b shows that, in sands, the major part of the compression under a certain increment of load takes place almost instantaneously. In this example, about 90 percent of the compression has occurred after the load has been acting for only 1 minute. Saturation of the sand sample will not alter this

relationship provided that the water content of the sand can change freely. This is usually the case in the field because of the high permeability of the sand.

3. From Fig. 6-1b it appears that the time lag during compression in sands is small. It is caused primarily by a shear-induced motion of sand particles that is frictional in nature. Therefore the time lag in reaching the final state in the sand is called the frictional lag.

4. Figure P-6.5b shows the relatively considerable time required for the occurrence of the compression in clays under a certain increment of load. The shape of this curve is different from that for sand (Fig. 6.1b) in that the curve for clay does not approach a horizontal asymptote but continues on a gentle slope, as indicated in Fig. P-6.5b by the dash line. The dash line represents the continuous nature of the deformation of the clay sample beyond the 24-hr period of time for the experiment, and is known as the secondary time effect. This type of consolidation is usually the case in the field, where buildings founded on thick-clay strata undergo settlements that continue for long periods of time (sometimes many years) at steadily decreasing rates (see Section6.4).

5. From point (4) it becomes clear that there is a large time lag in the consolidation of clays. This lag is attributed primarily to

    (a) Hydrodynamic lag, which is the time required for the water to escape from the voids under the imposed pressure and a function of the permeability of the soil. The hydrodynamic lag usually increases with the decrease of permeability.

    (b) Plastic lag, which is a complicated factor and a function of the plastic action in absorbed water near grain-to-grain contacts (see Section 6.4). The frictional lag in sands may be regarded as a special simple form of plastic lag.

Consolidation and Settlement

6.6 Derive the general equation expressing the compression of a confined stratum of normally consolidated clay with a thickness H as a result of the increase of effective pressure from p_0 to $p_0 + \Delta p$ on the clay stratum.

Solution From Table P-1.7, Chapter 1, the relationship between the void ratio e and the porosity n is

$$n = \frac{e}{1+e} \tag{a}$$

Thus the decrease in porosity Δn per unit of the original volume of soil, corresponding to a decrease in void ratio Δe, may be obtained by means of equation (a) as

$$\Delta n = \frac{\Delta e}{1 + e_0} \tag{b}$$

in which e_0 is the initial void ratio.

Substituting (6-2) into (b) will yield

$$\Delta n = \frac{a_v}{1 + e_0} \Delta p = m_v \Delta p \tag{c}$$

in which

$$m_v\,(\mathrm{cm^2/g}) = \frac{a_v\,(\mathrm{cm^2/g})}{1 + e_0} \tag{d}$$

where m_v is known as the coefficient of volume compressibility. It represents the compression of the clay, per unit of the original thickness, due to a unit increase of the pressure.

Then if H is thickness of a stratum of clay under an effective pressure p, an increase of the pressure from p_0 to $p_0 + \Delta p$ reduces the thickness of the stratum by

$$S = H \Delta p m_v \tag{e}$$

Combining equations (6-2) and (6-3) yields

$$a_v = \frac{C_c}{\Delta p} \log \frac{p_0 + \Delta p}{p_0} \tag{f}$$

Now, by combining (f) and (d), we obtain

$$m_v = \frac{C_c}{\Delta p (1 + e_0)} \log \frac{p_0 + \Delta p}{p_0} \tag{g}$$

Substituting the value of m_v as expressed in (e) into (g) yields

$$S = H \frac{C_c}{1 + e_0} \log \frac{p_0 + \Delta p}{p_0} \tag{h}$$

The relationship (h) is the general expression representing the compression S of a confined stratum of normally loaded ordinary clay.

6.7   A bed of sand 30 ft thick is underlain by a compressible stratum of clay 20 ft thick and having a liquid limit of 46 percent. The G.W.T. is at a depth of 15 ft below ground surface. The natural water content of the clay is 41 percent, and the specific gravity of the solid clay particles is 2.76. The submerged unit weight of the sand is 65 lb/ft^3, and the unit weight of the moist sand located above the G.W.T. is 110 lb/ft^3. Assuming that it has been established from geological evidence that the clay stratum is normally consolidated, estimate the average probable final settlement due to the construction of a building on the top of the sand that will cause an increase in the overburden effective pressure on the clay stratum by 1.1 t/ft^2.

The situation described above is illustrated in Fig. P-6.7. From Table P-1.7   *Solution* (Chapter 1), and for a saturated soil, we can determine the initial void ratio e_0 of the clay stratum as follows:

$$e_0 = w G_s = (0.41)(2.76) = 1.125$$

The submerged density of the clay γ_{sub} (from Table P-1.7) is

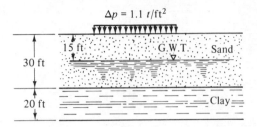

Figure P-6.7

$$\gamma_{sub} = \frac{(G_s - 1)\,\gamma_w}{1 + e_0} = \frac{(2.76 - 1)\,62.4}{1 + 1.125} = 52 \text{ lb/ft}^3$$

Effective pressure at the center of the clay layer, p_0, is

$$p_0 = (15)(110) + (15)(65) + (10)(52) = 3,470 \text{ lb/ft}^2 \text{ or } 1.735 \text{ t/ft}^2$$

Since the clay is normally consolidated, the value of C_c will be estimated from the approximate empirical relationship (6-5)

$$C_c = 0.009(11) - 10 = 0.009(46 - 10) = 0.324$$

Substitution of the previously calculated values of e_0, p_0, and C_c into equation (h) of Problem 6.6 will give the average settlement S as

$$S = H\left(\frac{C_c}{1 + e_0}\right)\log\left(\frac{p_0 + \Delta p}{p_0}\right) = 20\left(\frac{0.324}{1 + 1.125}\right)\log\left(\frac{1.735 + 1.1}{1.735}\right)$$

$$= 0.646 \text{ ft or } 7.8 \text{ in.}$$

6.8 A compressible clay stratum 20 ft thick has been consolidated under the weight of the overburden shown in Fig. P-6.8. On one side a valley was formed as a result of erosion and weathering processes. The original G. W. T. was originally at the ground surface; after the valley was formed it dropped to a level of 30 ft above the surface of the clay layer. A building is to be constructed on the valley floor, as shown in the figure; the result will be an increase of pressure on the clay stratum by 1.2 t/ft². The average liquid limit of the clay is 46 percent, its natural water content is 36 percent, and its specific gravity is 2.76. The overburden is a sand material that has the same unit weight as that in Problem 6.7

Compute : (a) the maximum consolidating pressure for the clay stratum under the valley floor (at A).

(b) the upper and lower limit for the settlement of the building shown Fig. P-6.8.

Solution (a) The original effective pressures p_0' on the surface of the clay layer before erosion of the valley is

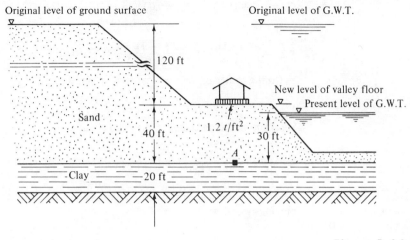

Original level of ground surface

Original level of G.W.T.

120 ft

Sand

New level of valley floor

Present level of G.W.T.

40 ft

1.2 t/ft²

30 ft

A

Clay

20 ft

$p_0' = (120 + 40)(65) = 10,400 \; \text{lb/ft}^2$

Existing effective pressure p_0 on the surface of the clay layer after erosion of the valley at A is

$p_0 = (10)(110) + (30)(65) = 3,050 \; \text{lb/ft}^2$

Therefore, the maximum consolidation pressure at A is

$p_0' - p_0 = 10,400 - 3,500 = 7,350 \; \text{lb/ft or } 3.675 \; t/\text{ft}^2$

(b) According to Terzaghi and Peck, the compressibility of preconsolidated clay depended on the ratio $\Delta p/(p_0' - p_0)$, in which Δp is the pressure added by the structure to the present overburden pressure p_0, (p_0 is the present overburden pressure, and p_0' is the maximum pressure that has ever acted on the clay or the preconsolidating pressure).

If this ratio is $< 60\%$, the compressibility of the clay is likely to be from 10 to 25 percent that of a similar clay in a normally consolidated state. With increasing values of the ratio, the effect of the precompression on the compressibility of the clay decreases. For ratio $> 100\%$, the influence of the precompression on the settlement of the structure can be disregarded.

Now, assuming that the clay stratum is normally consolidated instead of being preconsolidated, the settlement of the clay stratum under the pressure from the building can be calculated by using equation (h) of Problem 6.6, which is

$$S = H \frac{C_c}{1 + e_0} \log \frac{p_0 + \Delta p}{p_0} \qquad \text{(a)}$$

in which (for a clay stratum)

$$C_c = 0.009[(LL) - 10] = 0.009(46 - 10) = 0.324$$

$$e_0 = wG_s = (0.36)(2.76) = 0.9936$$

$$\gamma_{sub} = \frac{(G_s - 1)\gamma_w}{1 + e_0} = \frac{(2.76 - 1)62.4}{1 + 1.125} = 51 \text{ lb/ft}^3$$

p_0 (to the center of the clay layer) $= (15)(110) + (30)(65) + (10)(51)$
$$= 3,560 \text{ lb/ft}^2$$

Substituting the preceding values into equation (a) yields

$$S = (20)\left(\frac{0.324}{1 + 0.9936}\right) \log\left(\frac{1.780 + 1.2}{1.780}\right) = 0.74 \text{ ft or } 8.9 \text{ in.}$$

Using the empirical relationship developed by Terzaghi and Peck (and which was previously outlined),

$$\frac{p}{p_0' - p_0} = \frac{1.2}{3.675 - 1.780} 100 = \frac{1.2}{1.895} 100 \cong 64 \%$$

which is fairly close to the 60 percent limit previously discussed. Thus the upper and lower limits of settlement under the pressure from the structure would be

$$\Delta H_{min} = (0.10)(8.9) = (0.10)(8.9) = 0.89 \text{ in.}$$

$$\Delta H_{max} = (0.25)(8.9) = 2.5 \text{ in. [which is a more realistic value than the lower limit,}$$
since the ratio $\Delta p/(p_0' - p_0)$ is closer to the higher limit of 60 %]

6.9   A clay stratum 7 ft thick overlays an impervious bedrock. The clay stratum is topped by 17 ft of sand deposit. Ground water level is 7 ft below the surface of the ground. The wet density of the sand above the G. W. T. is 120 lb/ft³, the saturated density of the sand below G. W. T. is 132 lb/ft³, and the saturated density of the clay is 121 lb/ft³. The results of a consolidation test on an undisturbed sample of the clay, 1 in. thick and drained on both sides, is given in the following table:

TABLE P-6.9

Pressure (t/ft^2)	0.5	1.0	2.0	3.0	4.0	5.0
Void ratio	0.72	0.65	0.58	0.54	0.51	0.48

If a foundation was built on the surface of the ground, thus resulting in an additional pressure of 0.48 t/ft^2 at the center of the clay stratum, estimate the value of field compressibility.

Solution   Initial effective pressure at center of the clay stratum

$$= (7)(120) + (10)(132 - 62.4) + (3.5)(121 - 62.4) = 1751.1 \text{ lb/ft}^2$$

Total effective pressure at center of clay stratum

$$= 0.875 + 0.48 = 1.355 \ t/ft^2$$

If we plot the values of pressure against the void ratio from the consolidation test, Table P-6.9 will result in Fig. P-6.9.

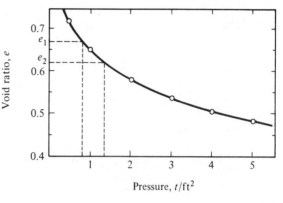

Figure P-6.9

From Fig. P-6.9, at a pressure $= 0.875 \ t/ft^2$, $e_1 = 0.67$; and at a pressure $= 1.355 \ t/ft^2$, $e_2 = 0.62$. By (6-7), the compressibility of the clay stratum is

$$\Delta H = \left(\frac{0.67 - 0.62}{1 + 0.67}\right)(84) = 2.5 \ in.$$

6.10 The following results were obtained in a consolidation test under an increment of load on a saturated clay sample. The initial thickness of the sample was 1 in. Drainage of the sample during test was from both sides. The thickness of sample after complete consolidation was 0.942 in. Determine the value of the coefficient of consolidation c_v.

TABLE P-6.10

Time t elapsed from application of pressure (minutes)	1	4	9	16	25	100	225	400
Compression in inches	0.0030	0.0050	0.0078	0.0105	0.0135	0.0255	0.0350	0.0410

Plotting the values of $\sqrt{t}$ against corresponding values of compression, as given *Solution* in Table P-6.10, will yield the curve in Fig. P-6.10. Extend the straight portion of the curve to cut 100 percent compression line *AB* and determine the corresponding

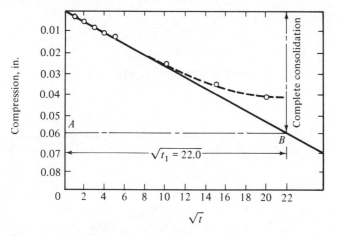

Figure P-6.10

value of $\sqrt{t_1}$. Referring to Fig. P-6.10,

$$\sqrt{t_1} = 22 \quad \text{or} \quad t_1 = 485 \text{ min.}$$

$$c_v = \frac{\pi H^2}{4t_1}$$

where H is the length of the drainage path. Since the sample drains from both sides in this case, $H = \frac{1}{2}$ thickness of sample = 0.5 in. Thus

$$c_v = \frac{\pi(0.5)^2}{(4)(485)} = 0.004 \text{ in.}^2/\text{min}$$

Consolidation

6.11   Using the assumptions introduced in Section 6.3, page 142, derive the governing differential equation of Terzaghi's consolidation theory (6-8).

Solution   The key point of this derivation is the assumption (3) calling for the equality of the rate of change in volume due to the pore-water outflow dQ/dt and the rate of change in volume due to the change in pore volume dV/dt.
Analytically,

$$\frac{dQ}{dt} = \frac{dV}{dt} \tag{a}$$

For an elemental soil cube $dx\ dy\ dz$ at the depth z below the surface, the amount of pore-water outflow (Problem 5.5) during the time interval dt is

$$dQ = \left(\frac{\partial v_x}{\partial x} + \frac{\partial v_y}{\partial y} + \frac{\partial v_z}{\partial z}\right) dx\ dy\ dz\ dt \tag{b}$$

which in the case of one-dimensional flow in the z direction (assumption 3) reduces to

$$dQ = \frac{\partial v_z}{\partial z} dx\, d\, y\, dz\, dt = k_z \frac{\partial^2 h}{\partial t^2} dx\, dy\, dz\, dt \tag{c}$$

where v_z = flow velocity

$\quad\quad k_z$ = permeability of soil

$\quad\quad h$ = hydraulic head

Simultaneously, the change in volume due to the change in pore volume of the soil cube during the time dt is

$$dV = \frac{\partial(V_s + eV_s)}{\partial t} dt = \frac{\partial e}{\partial t} V_s\, dt \tag{d}$$

where V_s = volume of solids (constant)

$\quad\quad eV_s$ = volume of voids (variable)

The substitution of (c) and (d) into (a) yields

$$k_z \frac{\partial^2 h}{\partial t^2} dx\, dy\, dz = \frac{\partial e}{\partial t} V_s \tag{e}$$

where

$$V_s = \frac{V}{1 + e_0} = \frac{dx\, dy\, dz}{1 + e_0} \tag{f}$$

and e_0 = initial void ratio

Thus

$$k_z \frac{\partial^2 h}{\partial z^2} = \frac{1}{1 + e_0} \frac{\partial e}{\partial t} \tag{g}$$

and since the flow is caused by the hydrostatic excess pressure

$$u = h\gamma_w$$

equation (g) can be expressed as

$$\frac{k_z}{\gamma_w} \frac{\partial^2 u}{\partial z^2} = \frac{1}{1 + e_0} \frac{\partial e}{\partial t} \tag{h}$$

Since by (6-2)

$$\partial e = -a_v\, dp = a_v\, du \tag{i}$$

where a_v = coefficient of compressibility, (h) can be written as

$$\frac{k_z(1 + e_0)}{a_v \gamma_w} \frac{\partial^2 u}{\partial z^2} = \frac{\partial u}{\partial t}$$

and in terms of (6-10) becomes

$$c_v \frac{\partial^2 u}{\partial t^2} = \frac{\partial u}{\partial t}$$

which is the desired equation (6-8).

6.12 Using the method of separation of variables, find the general integral of the consolidation equation (6-8).

Solution In order to find the integral of (6-8), a product function

$$u(z, t) = Z(z)T(t) \tag{a}$$

is assumed to be the solution.

Then (6-8) in terms of partial derivatives of (a) becomes

$$\frac{Z''}{Z} = \frac{\dot{T}}{c_v T} \tag{b}$$

where

$$Z'' = \frac{\partial^2 Z}{\partial z^2} \quad \text{and} \quad \dot{T} = \frac{\partial T}{\partial t}$$

Since the left side term in (b) is a function of z and the right side term is a function of t, each side must be equal to the same constant (say $\pm \lambda^2$). Choosing

$$\frac{Z''}{Z} = -\lambda^2 \quad \text{and} \quad \frac{\dot{T}}{T} = -\lambda^2 c_v \tag{c, d}$$

gives

$$Z'' + \lambda^2 Z = 0 \qquad \dot{T} + \lambda^2 c_v T = 0 \tag{e, f}$$

the solutions of which are

$$Z(z) = C_1 \cos \lambda z + C_2 \sin \lambda z \qquad T(t) = C_3 \exp(-\lambda^2 c_v t) \tag{g, h}$$

The product of (g, h) is then the general integral of (6-8), given as

$$u(z, t) = (C_1 \cos \lambda z + C_2 \sin \lambda z)C_3 \exp(-\lambda^2 c_v t) \tag{i}$$

or simply

$$u(z, t) = (C_4 \cos \lambda z + C_5 \sin \lambda z) \exp(-\lambda^2 c_v t) \tag{j}$$

where C_4, C_5 are the constants of integration and λ is the characteristic value.

For the determination of C_4, C_5, and λ, boundary conditions of Fig. 6-2 are available.*

6.13   A sample of clay 1 in. thick was tested in a consolidometer. If 50 percent of the primary consolidation of the sample takes place in the first 4 minutes, how long will it take a structure on a 10 ft stratum of the same clay to experience 50 percent of its final primary consolidation. Assume similar drainage conditions for both the laboratory specimen and the field stratum.

By (6-13), that is   *Solution*

$$T_v = \frac{c_v}{H^2} t \tag{a}$$

Since field and laboratory conditions are similar,

$$\frac{t_l}{t_s} = \frac{H_l^2}{H_s^2} \quad \text{or} \quad t_s = \frac{H_s^2}{H_l^2} t_l \tag{b}$$

where subscripts s and l refer to clay stratum in the field and clay specimen in the laboratory, respectively. Substituting in relationship (b) for $H_l = \frac{1}{2}$ thickness of sample $= 0.5$ in. and $H_s = \frac{1}{2}$ depth of field stratum $= 5$ ft or 60 in. yields

$$t_s = \frac{(60)^2}{(0.5)^2} \frac{4}{(60)(24)} = 40 \text{ days} \tag{c}$$

6.14   The vertical effective pressure on a stratum of normally consolidated clay 10 ft thick is increased from 2 to 3 t/ft^2. The stratum is confined between an impermeable layer at the bottom and sand layer at the top. Its compression index C_c is 0.21, its consolidation coefficient c_v is 0.018 $in.^2/min$, and its initial void ratio is 0.83. Calculate:
(a)   the compression due to consolidation.
(b)   the time required to attain 80 percent consolidation. (Assume uniform distribution of consolidating pressure.)

(a)   By (6-3)   *Solution*

$$\Delta e = C_c \log \frac{p_0 + \Delta p}{p_0} = 0.21 \log \frac{3}{2} = 0.0315$$

Then by (6-7)

$$H = \frac{H \Delta e}{1 + e_0} = \frac{(10)(0.0315)}{1 + 0.83} = 0.172 \text{ ft or } 2.06 \text{ in.}$$

(b)   For 80 percent consolidation,

$$T_v = 0.567 \quad \text{(Fig. 6-4, curve } C_1 \text{, or condition 1, Table 6-2)}$$

*D. W. Taylor, Soil Mechanics, Wiley, New York, 1948, pp. 229–34.

By (6-13)

$$T_v = \frac{c_v}{H^2} t \quad \text{or} \quad t = \frac{T_v H^2}{c_v} = \frac{(0.567)(10 \times 12)^2}{0.018} = 4.52 \times 10^5 \text{min or } 315 \text{ days}$$

6.15   In the clay stratum described in Problem 6.14, calculate the pore-water pressure and the vertical effective pressure at the middle of the clay stratum at time t determined in part (b) of the problem. Comment on the results.

Solution   If a confined clay stratum is draining on both sides (top and bottom), the value of u_i will fall instantaneously to zero at these upper and lower surfaces. If it is only draining on one side, such as in this problem, the value of u_i will fall to zero instantaneously at the free draining surface and will increase progressively toward the impervious layer. With a mathematical solution, it is possible to determine u at time t for any point within the layer. If these values of pore pressure are plotted, a curve known as an isochrone can be drawn through the points. The plot of isochrones for

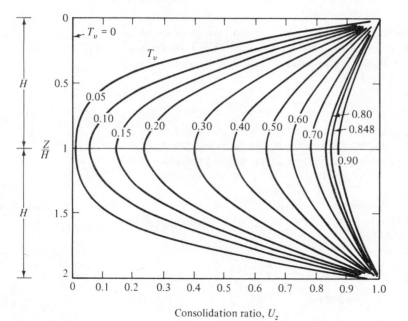

Figure P-6.15

different time intervals is shown in Fig. P-6.15. For a particular point at a depth z, the degree of consolidation U_z is

$$U_z = \frac{u_i - u_z}{u_i} \quad \text{or} \quad = \frac{(p_2 - p_1) - u_z}{p_2 - p_1} = \frac{\Delta p - u_z}{\Delta p}$$

In this problem, the stratum is free to drain on only one side, so only the lower or upper part of the symmetrical set of curves in Fig. P-6.15 will be used.

Since $T_v = 0.567$ [see part (b) of P-6.14a] and $z/H = 0.5$, the value of U_z is 0.77 (from Fig. P-6.15). The initial excess hydrostatic pore-water pressure u_i is

$$u_i = p_2 - p_1 = \Delta p = 3 - 2 = 1 \ t/ft^2 \quad \text{or} \quad 2{,}000 \ lb/ft^2$$

Hence

$$U_z = \frac{\Delta p - u_z}{\Delta p} \quad \text{or} \quad 0.77 = \frac{2{,}000 - u_z}{2{,}000}$$

Thus, the pore-water pressure at the middle of the stratum is

$$u_z = 460 \ lb/ft^2$$

The effective pressure $\bar{p}$ at any time t is $p - u$. After loading, the pressure on the stratum of clay was increased to $3 \ t/ft^2$. Thus at $t = 315$ days, the effective pressure $\bar{p}$ is

$$\bar{p} = (3 \times 2{,}000) - 460 = 5{,}540 \ lb/ft^2 \quad \text{or} \quad 2.77 \ t/ft^2$$

It can be observed that the value of U_z calculated in Problem 6.15 is not equal to the 0.80 given in Problem 6.14 for the stratum under the same conditions. This difference is attributed to the fact that the degree of consolidation is not uniform throughout the clay stratum. The value of 0.80 determined in Problem 6.14 is the average degree of consolidation for the entire layer, whereas the value of 0.77 in this problem is the degree of consolidation at the middle of the stratum.

Rheology of Consolidation

6.16 Discuss the basic concept of rheology of consolidation in connection with the application of linear models only.

The central idea of the linear rheological modeling is the construction of a model *Solution* whose time-dependent stress-strain relation, when inserted in the governing equation of consolidation, yields a response closely approximating the behavior of the soil system.

Symbolically, if the consolidation equation (6-8) is stated as

$$\frac{k_z}{\gamma_w} \frac{\partial^2 u}{\partial z^2} = -\frac{\partial \epsilon}{\partial t} \tag{a}$$

and if

$$\epsilon = (k_1, k_2, \ldots, \eta_1, \eta_2, \ldots, t) \tag{b}$$

$k_1, k_2, \ldots = $ elastic spring constants

$\eta_1, \eta_2, = $ coefficients of viscosity

are known, then with

$$\partial \epsilon = \partial \sigma(k_1, k_2, \ldots, \eta_1, \eta_2, \ldots t) = -\partial u \tag{c}$$

equation (a) becomes

$$\frac{k_z}{\gamma_w}\frac{\partial^2\sigma}{\partial z^2} = \frac{\partial\sigma}{\partial t} \tag{d}$$

and includes the elastic but also the viscoelastic effects.

6.17   Using the outline of Problem 6.16, discuss the rheological model used by Gibson and Lo* for the interpretation of consolidation of clay.

Solution   The rheological model used by these investigators consists of three elementary models (Section 3.2), two elastic springs of spring constant k_1, k_2, and one dashpot of coefficient of viscosity η (Fig. P-6.17).

The functioning of this model is described as follows:

1. The time dependent effective stress $\sigma(t)$ acting on the model at depth z below the ground compresses the spring 2 instantaneously (Hookean model) and the settlement thus produced corresponds to the elastic deformation of soil.

2. Because of the dashpot, the deformation of the remaining part of the model (Kelvin model) is delayed and becomes time dependent. This corresponds to the transfer of stress from the pore water to the soil skeleton.

3. The full load first taken fully by the dashpot (pore water) is progressively transferred to the spring 1.

4. After an appreciable amount of time has elapsed, the capacity of the dashpot is exhausted (drainage is completed) and the load is carried by 1 and 2 springs.

The transition between steps 3 and 4 is corresponding to the secondary consolidation.

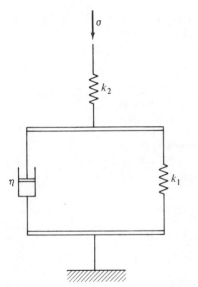

Figure P-6.17

6.18   Show the mathematical formulation of the Gibson-Lo experiment.

Solution   By definition of the model, the total strain is

$$\epsilon = \epsilon_H + \epsilon_K \tag{a}$$

where ϵ_H = the strain in the Hookean model

ϵ_K = the strain in the Kelvin model

For $\epsilon(0) = 0$, equation (a), written in terms of the results of Chapter 3, becomes

$$\epsilon = \frac{\sigma(t,z)}{k_2} + \frac{1}{\eta}\int_0^t \sigma(\tau,z)\exp\left[\frac{k_1(\tau - t)}{\eta}\right]d\tau \tag{b}$$

*R. E. Gibson and K. Y. Lo, *A Theory of Consolidation for Soils Exhibiting Secondary Compression*, Norwegian Geotechnical Institute, Oslo Publication No. 41, 1961.

where by assumption

$$d\epsilon \doteq \frac{-de}{1+e} \tag{c}$$

Then relating

$$\frac{k_z}{\gamma_w} \frac{\partial^2 u}{\partial z^2} = -\frac{\partial \epsilon}{\partial t} \tag{d}$$

$$\frac{k_z}{\gamma_w} \frac{\partial^2 u}{\partial z^2} = \frac{-\partial \left[\dfrac{\sigma(t,z)}{k_2} + \dfrac{1}{\eta} \displaystyle\int_0^t \sigma(\tau, z) \exp\left[\dfrac{k_1(\tau - t)}{\eta} \right] d\tau \right]}{\partial t} \tag{e}$$

and since $\partial u = -\partial \sigma(t, z)$ then (e) becomes

$$\frac{k_z}{\gamma_w} \frac{\partial^2 \sigma(t,z)}{\partial z^2} = \frac{1}{k_2} \frac{\partial \sigma(t,z)}{dt} + \frac{\sigma(t,z)}{\eta} - \frac{k_1}{\eta^2} \int_0^t \sigma(\tau, z) \exp\left[\frac{k_1(\tau - t)}{\eta} \right] d\tau \tag{f}$$

The solution of this equation requires the application of Laplace transforms and can be found in the reference cited in Problem 6.17.

Although the mathematics and the labor involved are far beyond the level of this book, Problems 6.16, 6.17, and 6.18 demonstrate how the rheological models work physically and mathematically.

SUPPLEMENTARY PROBLEMS

Immediate Elastic Deformation

6.19   A soil material was found to have the following characteristics: percent passing No. 200 sieve is 57, liquid limit (LL) is 47 percent, and plastic limit (PL) is 35 percent. If the soil is loaded with a rectangular slab 12 ft long and 9 ft wide and carrying a uniformly distributed load of 30 lb/in.^{2}, estimate the immediate elastic displacement of the surface of the soil mass.

This soil is considered as plastic clay (see Chapter 2, page 34). Assume that the value of $E_0 = 10,000$ lb/in.^{2} Thus displacement of the surface $\Delta_i = 0.618$ in.   *Answer*

Consolidation Settlement

6.20   A normally consolidated clay stratum 20 ft thick is drained only on its upper side. A consolidation test was performed on an undisturbed sample of the clay material, and the following results were obtained :

at $p_1 = 1.5$ kg/cm^2,

$e_1 = 1.41$

and at $p_2 = 2.0$ kg/cm^2,

$e_2 = 1.30$

The coefficient of permeability under this increment of loading was determined (see Problem 6.2) and found to be 15×10^{-8} cm/sec. Assuming that the preconsolidating pressure was 1.5 kg/cm² and that the stratum will be loaded to 2.4 kg/cm², compute

(a) the compression of the clay stratum in inches.

(b) settlement at $U = 50\%$.

(c) time in seconds to reach 50 % consolidation.

Answer (a) $\Delta H = 11$ in.

(b) ΔH at 50 % $U = 5.5$ in.

(c) t_{50} (to 50 % U) = 450 days

6.21 A buried, normally consolidated clay stratum 40 ft thick is drained from both sides. The stratum is overlaid by a sand deposit 40 ft thick and having a saturated unit weight of 130 lb/ft³. Its dry unit weight is 110 lb/ft³. The saturated unit weight of the clay is 120 lb/ft³, its void ratio e is 1.16, its coefficient of compressibility a_v is 0.06 ft²/ton, and its coefficient of consolidation c_v is 0.021 in.²/min. Assume that a 20-ft lowering of the water table occurred in the sand deposit overlying the clay. What settlement will eventually take place at ground surface because of the consolidation of the clay stratum due to lowering of the G. W. T., and how much time will be required for 60 percent of this settlement to occur?

Answer $\Delta p = 0.524\, t/\text{ft}^2$, $\Delta H = 6.95$ in., $t_{60} = 550$ days or 1.5 years

Consolidation Test

6.22 The following results (Table P-6.22) were obtained from a consolidation test on a clay sample, the properties of which are

Initial thickness of the sample	= 3.81 cm
Liquid limit (LL)	= 45 %
Specific gravity of particles	= 2.72
Diameter of sample	= 10.7 cm
Initial weight of consolidation test specimen	= 650 g
Oven-dry weight of specimen after test is completed	= 480 g

Each increment of load was allowed to remain for 24 hours; it was assumed that primary consolidation was complete at the end of this period.

TABLE P-6.22

Pressure (kg/cm²)	Dial Readings (mm × 10⁻²)	Pressure (kg/cm²)	Dial Readings (mm/10⁻²)
0.000	0	1.04	186
0.130	28	2.08	338
0.260	55	4.16	508
0.520	98	8.32	669

(a) Compute the void ratio at each increment of load and then plot the e-log p curve.
(b) Using Casagrande's graphical construction method (Problem 6.3), determine the preconsolidation pressure.
(c) Determine the compression index C_c by means of (6-5) and compare with the one obtained by using the results of the test.
(d) Compute the value of the coefficient of compressibility a_v and the coefficient of volume compressibility m_v for increment of pressure from 1.04 to 2.08 kg/cm².

(a) The e-log p curve is shown in Fig. P-6.22.   *Answer*

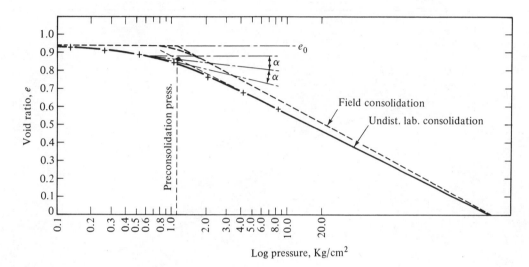

Log pressure, Kg/cm²

Figure P-6.22

(b) The preconsolidation pressure is determined on Fig. P-6.22, using Casagrande's graphical construction, and is 1.1 kg/cm².
(c) The compression index C_c from the test data is the slope of the straight-line portion of the field consolidation curve. From Fig. P-6.22, at $p_1 = 2$ kg/cm², $e_1 = 0.86$, and at $p_2 = 20$ kg/cm², $e_2 = 0.52$.

$$C_c = \frac{e_1 - e_2}{\log p_2 - \log p_1} = \frac{0.86 - 0.52}{1} = 0.340$$

From (6-5)

$$C_c = 0.009\,(LL - 10) = 0.315$$

(d) $a_v = 7.6 \times 10^{-5}$ cm²/gm, $m_v = 4.12 \times 10^{-5}$ cm²/g

6.23 Given the time-dial reading data as tabulated in Table P-6.23 resulting from a consolidation test of the same sample described in Problem 6.22, and under the pressure of 2.08 kg/cm².

(a) Plot the dial reading-time curve and determine the readings corresponding to 0 and 100 percent consolidation.

(b) Compute the value of k under this increment of load.

(c) If this sample was taken from a stratum 30 ft thick and drained from both sides (top and bottom), what is the time required for this stratum to reach 50 percent consolidation under a load of about 2.1 kg/cm²?

Answer (a) The dial reading-time curve is shown in Fig. P-6.23 and the 0 and 100 percent consolidation readings are marked (see Problem 6.4).

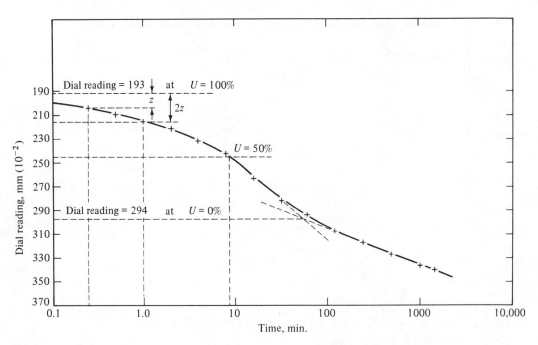

Figure P-6.23

(b) $c_v = 0.0013$ cm²/sec.

Note: The following relationship is used to calculate c_v from the dial reading-time curve:

$$c_v = \frac{T_v}{t} H^2.$$

Because of the secondary consolidation, the theoretical and experimental compression-time curves fit only up to about $U = 60\%$. Thus in determining the value of c_v, one must use values of t corresponding to about $U = 50\%$. For a sample drained on both sides (at $U = 50\%$), $T_v = 0.197$ (Table 6-2), and $H = \frac{1}{2}$ thickness of sample = 1.90 cm. $k = 5.7(10^{-8})$ cm/sec.

(c) Time for the stratum to reach 50 percent consolidation $= 345$ days.

Time, min	0.00	0.10	0.25	0.50	1.0	2.0	4.0	8.0	16.0	32.0	60.0
Dial Reading mm (10^{-2})	194.0	200.0	204.5	210.0	216.1	221.9	232.0	242.2	264.3	282.0	294.0

Time, min	120.0	240.0	480.0	1,000.0	1,440
Dial Reading mm (10^{-2})	308.2	317.9	327.0	336.0	400.0

stress
analysis
7

7.1 INTRODUCTION

General

The purpose of the stress and deformation analysis of soil systems is the determination of the stress and deformation field generated by the loads applied at the ground surface or within the subgrade. Because of the three-phase constitution, influence of time, changes in moisture, and variation of the soil structure (composition), the construction of a reliable mathematical model for a particular soil system is very difficult.

Elastic Model

For lack of a better approach (and also because of the plausible simplicity), the stress and deformation analysis of soil systems has been developed on the basis of the linear theory of elasticity, known as the *J. Boussinesq method* (1885). For all its deficiencies, and despite the crudeness of its results, this method remains the most commonly used tool of the half-space analysis and as such is presented in this chapter.

Assumptions

The following properties assigned to the soil system are introduced as the assumptions of Boussinesq's method:

1. The system (loads and soil) is in a state of static equilibrium.
2. All loads have been applied gradually and no kinetic energy is imparted.
3. The system is conservative and independent of time.
4. The soil is weightless, continuous, homogeneous, isotropic, and linearly elastic.
5. The material constants are known from experiments and are independent of time.

With the exception of (1) and (2), the remaining assumptions are not and cannot be satisfied by the real system. Yet in the absence of other more realistic approaches, this method is a valuable tool for the estimation of stresses at any depth beneath the surface.

Geometry

The geometry of a soil system is defined by the position vector of each point, the components of which are usually given in the cartesian or cylindrical coordinate system and are called the position coordinates.

In the cartesian system with origin i, the point j is given by x_{ij}, y_{ij}, z_{ij}; in the cylindrical system the same point is given by $r_{ij}, \omega_{ij}, z_{ij}$ (Fig. 7-1).

The signs of these coordinates are governed by the sign convention of geometry; that is,

1. All *coordinates are positive* if measured in the positive direction of coordinate axes.
2. All *angles are positive* if measured in the right-hand motion on the positive direction of coordinate axes.

Since the half-space extends downward, the Z axis in all figures points in that direction.

Static Vector

Loads acting on the soil surface are called *static causes*, and their *static effect* within the soil system is known as the stress. The stress in this chapter is defined as the intensity of internal force (force per unit area). As we know from the mechanics of solids, two types of stress occur in a continuum.

1. The *normal stress* ($\sigma_x, \sigma_y, \sigma_z,$ or $\sigma_r, \sigma_t, \sigma_z$) is perpendicular to the plane of its action and is denoted by the subscript of direction.
2. The *shearing stress* ($\sigma_{xy}, \sigma_{yz}, \sigma_{zx}, \ldots$ or $\sigma_{rt}, \sigma_{tz}, \sigma_{zr}, \ldots$) is tangential to the plane of its action and is denoted by the subscripts, the

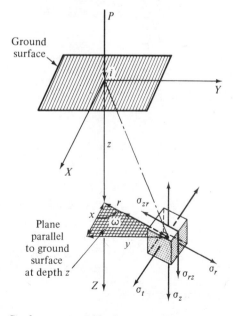

Single concentrated load normal to ground

Figure 7-1

first of which identifies the normal of the plane of action and the second of which is the subscript of direction.

The state of stress at a point is usually represented by an elemental free-body sketch, such as those shown in Figs. 7-1 to 7-4.

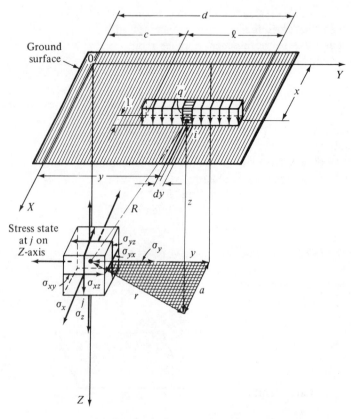

Figure 7-2   *Line load of constant intensity*

The signs of these stresses are governed by the sign convention of geometry —that is, all stresses acting on the far face of the element (away from the co-ordinate plane) in the direction of the parallel coordinate axis are positive (for the stresses acting on the near face, the opposite is true).

The signs of loads are governed by the sign convention of geometry; and since all loads in this chapter act downward, they are positive.

Displacement Vector

The deformation effect of loads is the *displacement*. In general, the displacement vector of a point j in the soil system is given by three components, δ_{jx}, δ_{jy}, δ_{jz}, of which usually only the vertical displacement is significant.

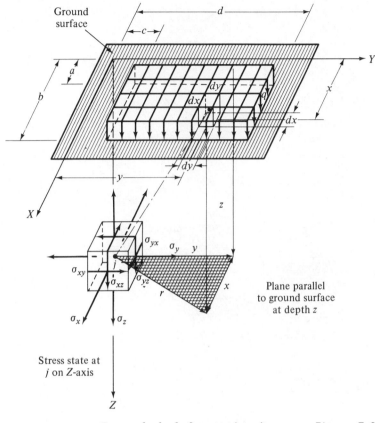

Rectangular load of constant intensity   **Figure 7-3**

The signs of displacements are governed by the sign convention of geometry, and in this chapter the downward displacements are positive.

Constants

In these calculations, two material constants are used.

1. Poisson's ratio μ, the value of which is usually taken as $\mu \cong 0.4 - 0.6$.
2. The modulus of subgrade E_0, the value of which is determined experimentally or is estimated by use of Table 6-1.

The conversion relations for E_0 are given as

$$E_0 = X \, \text{g/cm}^2 = 2.05X \, \text{lb/ft}^2 \quad E_0 = X \, \text{lb/ft}^2 = 0.488X \, \text{g/cm}^2$$

where X is a pure number.

   The average value of $\mu \cong 0.5$ is introduced in this chapter and is consis-
tently used in the development of all subsequent formulas.

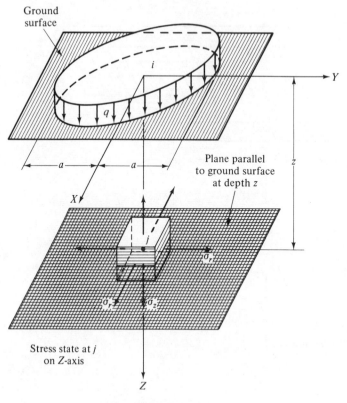

Ground
surface

i

q

Y

|← *a* →|← *a* →|

Plane parallel
to ground surface
at depth *z*

X

z

j

σ_r

σ_{rz} σ_z

Stress state at *j*
on Z-axis

Z

Figure 7-4   *Circular load of constant intensity*

7.2 ANALYTICAL METHOD

Single Concentrated Load

For the semi-infinite medium (half-space) defined above and acted on by a single concentrated load P normal to the ground surface at i (Fig. 7-1), the stresses in the cylindrical coordinate system at j (any point within the medium) are given as

$$\sigma_z = -\frac{3Pz^3}{2\pi(r^2 + z^2)^{5/2}} = -\frac{P}{z^2}K$$

$$\sigma_r = -\frac{3Pr^2z}{2\pi(r^2 + z^2)^{5/2}} = -\frac{Pr^2}{z^4}K \qquad (7\text{-}1)$$

$$\sigma_{zr} = \sigma_{rz} = -\frac{3Prz^2}{2\pi(r^2 + z^2)^{5/2}} = -\frac{Pr}{z^3}K$$

where σ_z, σ_r are the normal stresses along z, r, respectively, and $\sigma_{zr} = \sigma_{rz}$ are the shearing stresses along r, z, respectively, $\mu = 0.5, \sigma_t = 0$.

The constant K is a dimensionless factor given as

$$K = \frac{3z^5}{2\pi(r^2 + z^2)^{5/2}} = \frac{3}{2\pi}\left(\frac{z}{R}\right)^5 \tag{7-2}$$

where

$$r = \sqrt{x^2 + y^2} \quad \text{and} \quad R = \sqrt{x^2 + y^2 + z^2}$$

The numerical values of K vary from 0.4775 for $r/z = 0.00$ to 0.0001 for $r/z = 6.15$, as shown in Table 7-1. The derivations of (7-1) and their applications are given in Problems 7.1 and 7.2 respectively.

Numerical Values of K   **TABLE 7-1**

$\dfrac{r}{z}$	K	$\dfrac{r}{z}$	K	$\dfrac{r}{z}$	K	$\dfrac{r}{z}$	K	$\dfrac{r}{z}$	K	$\dfrac{r}{z}$	K
0.00	0.4775	0.50	0.2733	1.00	0.0344	1.50	0.0251	2.00	0.0085	2.50	0.0034
0.02	0.4770	0.52	0.2625	1.02	0.0803	1.52	0.0240	2.02	0.0082	2.52	0.0033
0.04	0.4765	0.54	0.2518	1.04	0.0764	1.54	0.0229	2.04	0.0079	2.54	0.0032
0.06	0.4723	0.56	0.2414	1.06	0.0727	1.56	0.0219	2.06	0.0076	2·56	0.0031
0.08	0.4699	0.58	0.2313	1.08	0.0691	1.58	0.0209	2.08	0.0073	2.58	0.0030
0.10	0.4657	0.60	0.2214	1.10	0.0658	1.60	0.0200	2.10	0.0070	2.60	0.0029
0.12	0.4607	0.62	0.2117	1.12	0.0626	1.62	0.0191	2.12	0.0068	2.62	0.0028
0.14	0.4548	0.64	0.2040	1.14	0.0595	1.64	0.0183	2.14	0.0065	2.64	0.0027
0.16	0.4482	0.66	0.1934	1.16	0.0567	1.66	0.0175	2.16	0.0063	2.66	0.0026
0.18	0.4409	0.68	0.1846	1.18	0.0539	1.68	0.0167	2.18	0.0060	2.68	0.0025
0.20	0.4329	0.70	0.1762	1.20	0.0513	1.70	0.0160	2.20	0.0058	2.70	0.0024
0.22	0.4242	0.72	0.1681	1.22	0.0489	1.72	0.0153	2.22	0.0056	2.72	0.0023
0.24	0.4151	0.74	0.1603	1.24	0.0466	1.74	0.0147	2.24	0.0054	2.74	0.0023
0.26	0.4054	0.76	0.1527	1.26	0.0443	1.76	0.0141	2.26	0.0052	2.76	0.0022
0.28	0·3954	0.78	0.1455	1.28	0.0422	1.78	0.0135	2.28	0.0050	2.78	0.0021
0.30	0.3849	0.80	0.1386	1.30	0.0402	1.80	0.0129	2.30	0.0048	2.80	0.0021
0.32	0.3742	0.82	0.1320	1.32	0.0384	1.82	0.0124	2.32	0.0047	2.84	0.0019
0.34	0.3632	0.84	0.1257	1.34	0.0365	1.84	0·0119	2.34	0.0045	2.91	0.0017
0.36	0.3521	0.86	0.1196	1.36	0.0348	1.86	0.0114	2.36	0.0043	2.99	0.0015
0.38	0.3408	0.88	0.1138	1.38	0.0332	1.88	0.0109	2.38	0.0042	3.08	0.0013
0.40	0.3294	0.90	0.1083	1.40	0.0317	1.90	0.0105	2.40	0.0040	3.19	0.0011
0·42	0.3181	0.92	0.1031	1.42	0.0302	1.92	0.0101	2.42	0.0039	3.31	0.0009
0.44	0.3068	0.94	0.0981	1.44	0.0288	1.94	0.0097	2.44	0.0038	3.50	0.0007
0.46	0.2955	0.96	0.0933	1.46	0.0275	1.96	0.0093	2.46	0.0036	3.75	0.0005
0.48	0.2843	0.98	0.0887	1.48	0.0263	1.98	0.0089	2.48	0.0035	4.13	0.0003

As in any stress field problem, the lines connecting all points of equal stress below the surface of the ground are called the *isobars*. They form, in a three-dimensional problem, isopressure surfaces (bulbs), as illustrated in Problems 7.3, 7.4, and 7.5.

The stress equations (7-1) in the cartesian system are

$$\sigma_x = -\frac{3Px^2z}{2\pi R^5} = -\frac{Px^2}{z^4}K \qquad \sigma_{xy} = -\frac{3Pxyz}{2\pi R^5} = -\frac{Pxy}{z^4}K$$

$$\sigma_y = -\frac{3Py^2z}{2\pi R^5} = -\frac{Py^2}{z^4}K \qquad \sigma_{yz} = -\frac{3Pyz^2}{2\pi R^5} = -\frac{Py}{z^3}K \qquad (7\text{-}3)$$

$$\sigma_z = -\frac{3Pz^3}{2\pi R^5} = -\frac{P}{z^2}K \qquad \sigma_{zx} = -\frac{3Pxz^2}{2\pi R^5} = -\frac{Px}{z^3}K$$

where K is given by (7-2).

Line Load of Constant Intensity

For the same medium acted upon by a line load of intensity q and length l along the Y axis (Fig. 7-2) at $x = a$, the stresses at j in the cartesian system are obtained by integrating (7-3) in limits c, d for $P = qdy$.

The vertical stress σ_z at j is

$$\sigma_z = -\frac{3qz^3}{2\pi}\int_c^d \frac{dy}{R^5} = -\frac{3qz^3}{4\pi r_a^4}\left(\frac{d}{R_{ad}} - \frac{c}{R_{ac}} - \frac{d^3}{3R_{ad}^3} + \frac{c^3}{3R_{ac}^3}\right) \qquad (7\text{-}4)$$

where

$$R = \sqrt{x^2 + y^2 + z^2}, \quad r_a = \sqrt{a^2 + z^2}, \quad R_{ac} = \sqrt{a^2 + c^2 + z^2},$$
$$R_{ad} = \sqrt{a^2 + d^2 + z^2}$$

With the increasing availability of digital computers, the numerical evaluation of the integrals (7-3) is more convenient than the lengthy closed-form evaluation given above (Problem 7.6).

Rectangular Load of Constant Intensity

The general case arises when the surface load is a rectangular load of constant intensity q acting on an area $(b - a) \cdot (d - c)$, the sides of which are parallel to the X and Y axes respectively (Fig. 7-3).

The stresses at j in the cartesian system are obtained by double integration of (7-3) in limits a, b and c, d with $P = qdxdy$.

The vertical stress σ_z is

$$\sigma_z = -\frac{3qz^3}{2\pi} \int_a^b dx \int_c^d \frac{dy}{R^5}$$

$$= -\frac{q}{2\pi} \left[\frac{az}{r_a^2} \left(\frac{R_{ac}}{c} - \frac{R_{ad}}{d} \right) - az^3 \left(\frac{1}{cr_c^2 R_{ac}} - \frac{1}{dr_d^2 R_{ad}} \right) \right.$$

$$- \frac{bz}{r_b^2} \left(\frac{R_{ac}}{c} - \frac{R_{bd}}{d} \right) + bz^3 \left(\frac{1}{cr_c^2 R_{bc}} - \frac{1}{dr_d^2 R_{bd}} \right)$$

$$\left. + \tan^{-1} \frac{ac}{zR_{ac}} - \tan^{-1} \frac{ad}{zR_{ad}} - \tan^{-1} \frac{bc}{zR_{bc}} + \tan^{-1} \frac{bd}{zR_{bd}} \right] \qquad (7\text{-}5)$$

where the equivalents r and R are

$$r_a = \sqrt{a^2 + z^2} \qquad r_b = \sqrt{b^2 + z^2}, \dots$$
$$R_{ac} = \sqrt{a^2 + c^2 + z^2} \qquad R_{ad} = \sqrt{a^2 + d^2 + z^2}, \dots$$

as shown in Problem 7.7.

Circular Load of Constant Intensity

Finally, a special case occurs when a circular load of radius a and constant intensity q is applied on the ground surface (Fig. 7-4). The stresses in cylindrical coordinates on the Z axis are then

$$\sigma_z = q \left(\frac{z^3}{r_a^3} - 1 \right)$$

$$\sigma_r = \frac{q}{2} \left(3 \frac{z}{r_a} - \frac{z^3}{r_a^3} - 2 \right) \qquad (7\text{-}6)$$

where $r_a = \sqrt{a^2 + z^2}$.

At $z = 0$,

$$\sigma_z = \sigma_r = -q \qquad (7\text{-}7)$$

The maximum shearing stress occurs at the depth

$$z = 0.7079 \qquad (7\text{-}8)$$

on the Z axis and is $q/2\sqrt{3}$.

Displacements

The vertical displacements of the soil system due to causes introduced in the preceding articles of this section are given analytically in Table 7-2. The moduli of subgrade E_0 in these formulas are those given in Table 6-1 (Problem 7.8).

TABLE 7-2   *Vertical Displacements*

Symbols	Equivalents
δ_{jz} = vertical displacement at j	$R = \sqrt{x^2 + y^2 + z^2} = \sqrt{r^2 + z^2}$
E_0 = modulus of subgrade (Table 6-1)	$R_{ac} = \sqrt{a^2 + c^2 + z^2} = R_{ca}$
P = magnitude of concentrated load	$R_{bc} = \sqrt{b^2 + c^2 + z^2} = R_{cb}$
q = intensity of distributed load	$R_{ad} = \sqrt{a^2 + d^2 + z^2} = R_{da}$
π = Ludolf's number = 3.14159...	$R_{bd} = \sqrt{b^2 + d^2 + z^2} = R_{db}$

1. Single Concentrated Load P (Fig. 7-1)

$$\delta_{jz} = \frac{3P(R^2 + z^2)}{2\pi E_0 R^3}$$

2. Line Load of Constant Intensity q (Fig. 7-2).

$$\delta_{jz} = \frac{3q(d-c)}{4\pi E_0} \ln \frac{a + R_{ad}}{a + R_{ac}}$$

3. Rectangular Load of Constant Intensity q (Fig. 7-3).

$$\delta_{jz} = \frac{3q}{4\pi E_0} \left(a \ln \frac{c + R_{ac}}{d + R_{ad}} - b \ln \frac{c + R_{bc}}{d + R_{bd}} \right.$$
$$\left. + c \ln \frac{a + R_{ca}}{b + R_{cb}} - d \ln \frac{a + R_{da}}{b + R_{db}} \right)$$

4. Circular Load of Constant Intensity q (Fig. 7-4).

$$\delta_{jz} = \frac{3qa^2}{2E_0 \sqrt{a^2 + z^2}}$$

7.3 INFLUENCE FUNCTIONS

Concept

By definition, the *influence surface* (generalization of influence line) is the graphical representation of the variation of σ or δ at j due to a moving unit force acting at i. Thus i is a *sliding position* of the unit cause and j is *fixed position* where the effect is measured.

The *influence function* is the analytical representation of the influence surface. Since j is fixed, it is taken as the origin, and the influence function (hence also the influence surface) is expressed in terms of x_{ji}, y_{ji}, z_{ji}, where the first two coordinates are variables and z_{ji} is fixed.

Applications

With notation

$$S_{ji} = S(z_{ji}, r_{ji}) = -\frac{K_{ji}}{z_{ji}^4} \tag{7-9}$$

the stress influence values by (7-3) become

$$\sigma_{jix} = x_{ji}^2 S_{ji}, \quad \sigma_{jiy} = y_{ji}^2 S_{ji}, \quad \sigma_{jiz} = z_{ji}^2 S_{ji}$$

$$\sigma_{jixy} = x_{ji} y_{ji} S_{ji}, \quad \sigma_{jiyz} = y_{ji} z_{ji} S_{ji}, \quad \sigma_{jizx} = z_{ji} x_{ji} S_{ji} \tag{7-10}$$

where K_{ji} is given by (7-2) and S_{ji} is the stress influence factor.

The effect of a system of vertical loads at j is then obtained by superposition of the singular effects. For example,

$$\sigma_{jix} = x_{j1}^2 S_{j1} P_1 + x_{j2}^2 S_{j2} P_2 + \cdots + x_{jm}^2 S_{jm} P_m \tag{7-11}$$

where $i = 1, 2, \ldots, m$ (Problem 7.9).

Similar is the derivation of the displacement influence functions. If only the vertical displacement is required, the superposition of the formulas in Table 7-2, Case 1, gives the direct answer.

7.4 SHAPE FUNCTIONS

Concept

If instead of a system of concentrated loads, a line load or an area load is involved, the concept of the *line* or *area shape function* offers some advantages.

In the case of the line load of Fig. 7-2, the load is translated along the Y axis to make $c = 0$ and $d = l$. Then by (7-4) with $a/z = m$ and $l/z = n$, the vertical stress under the origin 0 at depth z (Table 7-3) becomes

$$\sigma_z = -\frac{q/z}{2\pi(1 + m^2)^2} \left[\frac{3n}{\sqrt{1 + m^2 + n^2}} - \left(\frac{n}{\sqrt{1 + m^2 + n^2}} \right)^3 \right]$$

$$= -\frac{q}{z} \psi_z(m, n) \tag{7-12}$$

where $\psi_z(m, n)$ is the shape function of the dimensionless parameters m, n. The chart of $\psi_z(m, n)$ is given in Table 7-3.

In the case of the rectangular load of Fig. 7-3, the load is translated along the X and Y axes to make $a = 0, b = u, c = 0$, and $d = v$. Then by (7-5) with $u/z = m, v/z = n$, the vertical stress under the indicated corner at depth z (Table 7-4) is

$$\sigma_z = -\frac{q}{2\pi} \left[\frac{mn(2 + m^2 + n^2)}{(1 + m^2)(1 + n^2)\sqrt{1 + m^2 + n^2}} + \tan^{-1} \frac{mn}{\sqrt{1 + m^2 + n^2}} \right]$$

$$= q\phi_z(m, n) \tag{7-13}$$

where $\phi_z(m, n)$ is the shape function of the dimensionless parameters m, n, the chart of which is given in Table 7-4.

TABLE 7-3   *Shape Function ψ_z, Line Load of Constant Intensity q*

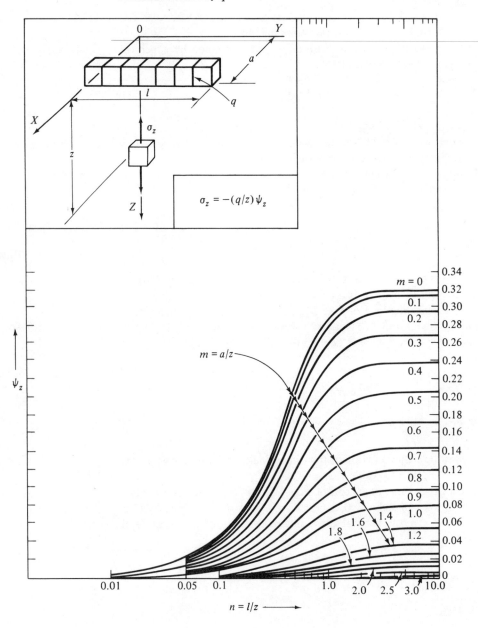

$$\sigma_z = -(q/z)\,\psi_z$$

R. E. Fadum, Influence Values for Estimating Stresses in Elastic Foundations, *Proceedings Second International Conference on Soil Mechanics and Foundation Engineering, Rotterdam, 1948, Vol. 3.*

**Shape Function ϕ_z, Rectangular
Load of Constant Intensity q**

TABLE 7-4

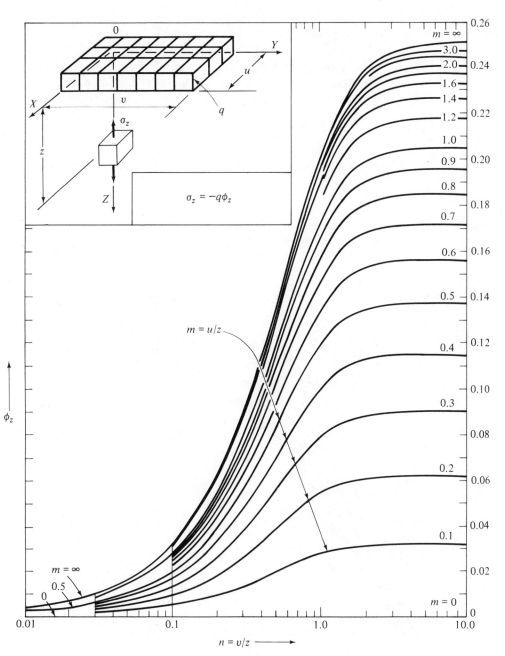

$\sigma_z = -q\phi_z$

$m = u/z$

$n = v/z \longrightarrow$

R. E. Fadum, Influence Values for Estimating Stresses in Elastic Foundations, *Proceedings
Second International Conference on Soil Mechanics and Foundation Engineering, Rotterdam, 1948, Vol. 3.*

Finally, in the case of the circular load of Fig. 7-4, the vertical stress beneath the center of the load at depth z is

$$\sigma_z = q\left[\left(\frac{m}{\sqrt{1+m^2}}\right)^3 - 1\right] = -q\rho_z(m, 0) \tag{7-14}$$

where $\rho_z(m, 0)$ is the shape function of the dimensionless parameter $m = z/a$. The chart of Table 7-5 can be used for the computation of σ_z at depth z under the centerline of the area or at any offset distance r. In this chart.

$$\rho_z(m, n) = \frac{-\sigma_z}{q} \tag{7-15}$$

where $m = z/a$ and $n = r/a$ are the dimensionless position parameters (Problems 7.10, 7.11, 7.12).

TABLE 7-5   *Shape Function ρ_z, Circular Load of Constant Intensity q*

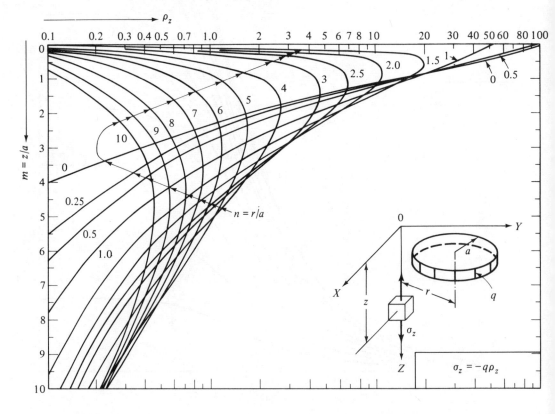

C. R. Foster and R. G. Ahlvin, Stresses and Deflections Induced by a Uniform Circular Load, *Proceedings Highway Research Board, Vol. 34, 1954, p. 320.*

The shape functions of Tables 7-3, 7-4, and 7-5 can be used directly for the determination of σ_z at j due to area loads indicated in the tables or for the calculation of σ_z at any point due to the same loads. For the latter, the principle of addition and subtraction is used.

In the case of Fig. 7-5, the stress at k is

$$\sigma_z = -q[\psi_1(m, n_1) + \psi_2(m, n_2)] \tag{7-16}$$

where

$$m = \frac{a}{z}, \quad n_1 = \frac{b}{z}, \quad n_2 = \frac{c}{z}$$

and ψ_1, ψ_2 are taken from Table 7-3 (Problem 7-13).

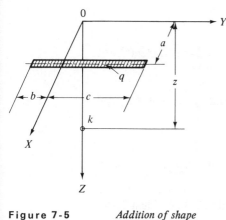

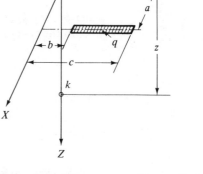

| Figure 7-5 | *Addition of shape functions ψ_z* | *Subtraction of shape functions ψ_z* | Figure 7-6 |

In the case of Fig. 7-6, the stress at k is

$$\sigma_z = -q[\psi_2(m, n_2) - \psi_1(m, n_1)] \tag{7-17}$$

and $m, n_1, n_2, \psi_1, \psi_2$ are the same values as in (7-16) (Problem 7.14).
    In the case of Fig. 7-7, the stress at k is

$$\sigma_z = -q[\phi_{11}(m_1, n_1) + \phi_{12}(m_1, n_2) + \phi_{21}(m_2, n_2) + \phi_{22}(m_2, n_2)] \tag{7-18}$$

where

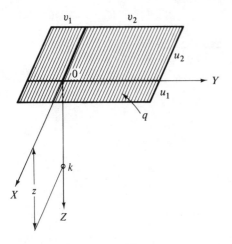

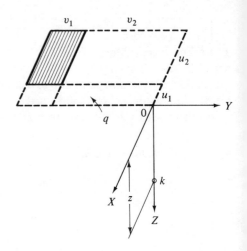

Figure 7-7   *Addition of shape functions ϕ_z*

Subtraction of shape functions ϕ_z   **Figure 7-8**

$$m_1 = \frac{u_1}{z}, \quad m_2 = \frac{u_2}{z}, \quad n_1 = \frac{v_1}{z}, \quad n_2 = \frac{v_2}{z}$$

and $\phi_{11}, \phi_{12}, \phi_{21}, \phi_{22}$ are taken from Table 7-4 (Problem 7.15).

In the case of Fig. 7-8, the stress at k is

$$\sigma_z = -q[\phi_{33}(m_3, n_3) - \phi_{23}(m_2, n_3) - \phi_{31}(m_3, n_1) + \phi_{21}(m_2, n_1)] \tag{7-19}$$

where

$$m_3 = \frac{u_1 + u_2}{z}, \quad n_3 = \frac{v_1 + v_2}{z}, \quad m_2 = \frac{u_2}{z}, \quad n_1 = \frac{v_1}{z}$$

and $\phi_{33}, \phi_{23}, \phi_{31}, \phi_{21}$ are taken from Table 7-4 (Problem 7.16).

7.5 NEWMARK'S INFLUENCE CHART

Concept

N. M. Newmark (1942) developed an *influence chart* that permits the integration of (7-1) graphically. This procedure is important because of its applicability to loaded areas of irregular shapes. Figure 7-9 shows the influence chart, which consists of influence areas of equal magnitude bounded by radial lines and arcs of circles. Each influence area represents a certain influence value i, which for the chart in Fig. 7-9 is $= 0.001$. Distance OQ on the chart represents the depth z below the ground surface at which the vertical stress is required.

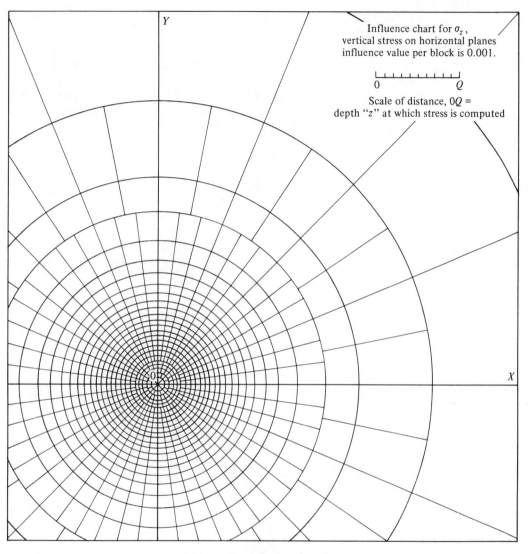

Newmark's influence chart for σ_z

N. M. Newmark, "Influence Charts for Computation of Stresses in Elastic Foundations," University of Illinois, *Eng. Exp. Station Bulletin 338*, 1947, p. 78.

Figure 7-9

Procedure

In order to use the chart, the plan of the loaded area is drawn on a tracing paper to the same scale as the depth z (at which stress calculation is required) equals the distance OQ on the chart. Then the point on the plan, or outside the plan of the loaded area at which it is required to find the vertical stress, is placed to coincide with the center of the chart. The number N of influence areas covered by the plan of loaded area are counted through the tracing paper. One may notice that the plan of the loaded area may be rotated through any angle about the axis, passing through the point where the stress is determined and the center of the chart without any change in the number of computed influence areas. Partial areas covered by the plan must be estimated and included in the total number counted.

Stress Equation

The vertical stress σ_z at the depth z is then given as

$$\sigma_z = Niq \qquad (7\text{-}20)$$

The application of Newmark's influence chart in conjunction with (7-20) is shown in Problem 7-17 as an illustration of a procedure widely used in engineering.

SOLVED PROBLEMS

Analytical Methods

7.1 Derive the stress equations (7-1) in the cylindrical coordinate system at j in the elastic semi-infinite medium (half-space) acted on by a concentrated load P normal to the ground surface at i as shown in Fig. 7-1.

Solution The general stress equations for an axially symmetrical case in the cylindrical coordinates known from the mathematical theory of elasticity* are

$$\sigma_z = \frac{\partial}{\partial z}\left[(2 - \mu)\nabla^2\phi - \frac{\partial^2\phi}{\partial z^2}\right]$$

$$\sigma_r = \frac{\partial}{\partial z}\left(\mu\,\nabla^2\phi - \frac{\partial^2\phi}{\partial z^2}\right)$$

$$\sigma_t = \frac{\partial}{\partial z}\left(\mu\,\nabla^2\phi - \frac{1}{r}\frac{\partial\phi}{\partial r}\right) \qquad (1)$$

$$\sigma_{zr} = \sigma_{rz} = \frac{\partial}{\partial r}(1 - \mu)\,\nabla^2\phi - \frac{\partial^2\phi}{\partial z^2}$$

where $\sigma_z, \sigma_r, \sigma_t, \sigma_{zr} = \sigma_{rz}$ are the stresses shown in Fig. 7-1, μ is Poisson's ratio of soil, ϕ is the stress function, and z, r are the coordinates.

The differential operator is

$$\nabla^2 = \frac{\partial^2}{\partial r^2} + \frac{1}{r}\frac{\partial}{\partial r} + \frac{\partial^2}{\partial z^2} \qquad (2)$$

---

*S. Timoshenko and J. N. Goodier, *Theory of Elasticity*, McGraw-Hill, New York, 2nd ed., 1951, p. 343.

and the stress function must satisfy

$$\nabla^2\nabla^2\phi = 0 \tag{3}$$

The solution* of (3) is

$$\phi = AR + Bz \ln\frac{R - z}{R + z} + Cz \ln r \tag{4}$$

where

$$R = \sqrt{r^2 + z^2}$$

and A, B, C are constants to be determined from the boundary conditions.

The substitution of (4) into (1) in terms of $\mu = 0.5$ (taken for soil as a good approximation) yields

$$\sigma_z = -\frac{(3A - 4B)z}{R^3} + \frac{3(A - 2B)r^2 z}{R^5}$$

$$\sigma_r = \frac{[(10B - 3A)r^2 + 8Bz^2]z}{r^2 R^3} + \frac{[3(A - 4B)r^2 - 6Bz^2]z^3}{r^2 R^5} + \frac{C}{r^2}$$

$$\sigma_t = -\frac{2Bz}{r^2 R} - \frac{C}{r^2} \tag{5}$$

$$\sigma_{zr} = \sigma_{rz} = -\frac{(3Az^2 + 2Br^2 - 2Bz^2)r}{R^5}$$

The constants A, B, C are computed from the conditions

$$\sigma_{zr}(z = 0) = 0, \quad P + \int_0^\infty \sigma_{zr} 2\pi r\, dz = 0, \quad \sigma_r(r = 0) = \sigma_t(r = 0) \tag{6}$$

and are

$$A = \frac{P}{2\pi}, \quad B = 0, \quad C = 0 \tag{7}$$

In terms of these values, (5) reduces to the final form (7-1) and $\sigma_t = 0$.

From Fig. 7-1, the stress equations (7-3) in the cartesian system are

$$\sigma_x = \sigma_r \cos^2 \omega = \left(-\frac{Pr^2 K}{z^4}\right)\left(\frac{x^2}{r^2}\right) = -\frac{Px^2}{z^4}K$$

$$\sigma_y = \sigma_r \sin^2 \omega = \left(-\frac{Pr^2 K}{z^4}\right)\left(\frac{y^2}{r^2}\right) = -\frac{Py^2}{z^4}K$$

$$\sigma_z = \sigma_z = -\frac{P}{z^2}K$$

*J. Boussinesq, *Application des potentiels a l'etude de l'equilibre et du movement des solide élastiques,* Mem. Soc. Sci. Agric. Arts, Lille, 1885, p. 212; reprinted by Gauthier-Villars, Paris, 1885.

$$\sigma_{xy} = \sigma_r \sin \omega \cos \omega = \left(-\frac{Pr^2K}{z^4}\right)\left(\frac{xy}{r^2}\right) = -\frac{Pxy}{z^4}K$$

$$\sigma_{yz} = \sigma_{rz} \sin \omega = \left(-\frac{PrK}{z^3}\right)\left(\frac{y}{r}\right) = -\frac{Py}{r^3}K$$

$$\sigma_{zx} = \sigma_{rz} \cos \omega = \left(-\frac{PrK}{z^3}\right)\left(\frac{x}{r}\right) = -\frac{Px}{z^3}K$$

where K is given by (7-2).

7.2 A single concentrated load $P = 100$ tons acts vertically on the surface of a uniform semi-infinite soil mass. Determine the value of the vertical stress σ_z at a point 10 ft below ground surface and 5 ft away from the line of action of the load.

Solution By (7-1)

$$\sigma_z = -\frac{P}{z^2}K$$

where $P = 100t$

$$z = 10 \text{ ft}$$

$$r = 5 \text{ ft}$$

By (7-2)

$$K = \frac{3z^5}{2\pi(r^2 + z^2)^{5/2}} = 0.2733$$

which must correspond to K given in Table 7-1 for $r/z = 5/10 = 0.5$.
In terms of these values, (7-1) yields

$$\sigma_z = -\left(\frac{100}{10^2}\right)0.2733 = -0.2733t/\text{ft}^2 = -546.6 \text{ lb/ft}^2$$

7.3 Construct the bulb of pressure (isopressure surface) connecting all points of vertical normal stress $\sigma_z = 0.2t/\text{ft}^2$ in Problem 7.2.

Solution From (7-1)

$$K = -\frac{\sigma_z z^2}{P} = \frac{0.2z^2}{100} = \frac{z^2}{500}$$

Since K is given by (7-2), the equation of the bulb of pressure $\sigma_z = -0.2t/\text{ft}^2$ is

$$\frac{3z^5}{2\pi(r^2 + z^2)^{5/2}} = \frac{z^2}{500}$$

or simply

$$750z^3 = \pi(r^2 + z^2)^{5/2}$$

The plotting of this equation can be done by computer or in tabular form, as shown in Table P-7.3.

Depth z (ft)	0.5	1.0	2.0	3.0	4.0	5.0	6.0	7.0	8.0	9.0	10.0	15	15.4
$K = z^2/500$	0.005	0.002	0.008	0.018	0.032	0.050	0.072	0.098	0.128	0.162	0.200	0.450	0.478
r/z (from Table 7-1)	3.750	2.840	2.035	1.650	1.400	1.210	1.060	0.940	0.830	0.735	0.650	0.155	0.000
r (ft)	1.87	2.84	4.06	4.95	5.60	6.05	6.36	6.58	6.64	6.61	6.50	2.33	0.00
σ_z (t/ft²)	−0.2	−0.2	−0.2	−0.2	−0.2	−0.2	−0.2	−0.2	−0.2	−0.2	−0.2	−0.2	−0.2

These values of r and z in Table P-7.3 give the coordinates of the points of σ_z of $0.2t/\text{ft}^2$. Plotting these coordinates (Fig. P-7.3) and then joining all of them with a smooth curve result in the isopressure surface of $0.2t/\text{ft}^2$.

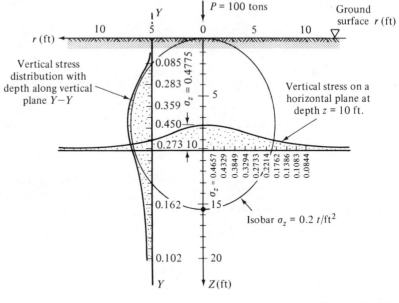

Figure P-7.3

7.4   For Problem 7.3, show the variation of the vertical stress σ_z on the horizontal plane at $z = 10$ ft below the ground surface.

Solution By (7-1)

$$\sigma_z = -K\frac{P}{z^2} = -K\frac{100}{10^2} = -K$$

where by (7-2)

$$K = \frac{3z^5}{2\pi(r^2 + z^2)^{5/2}} = \frac{300,000}{2\pi(r^2 + 100)^{5/2}}$$

Then the desired variation of σ_z becomes a function of r; that is,

$$\sigma_z = -\frac{300,000}{2\pi(r^2 + 100)^{5/2}}$$

as shown in Fig. P-7.3.

7.5 For Problem 7.3, show the variation of the vertical stress σ_z at $r = 5$ ft.

Solution By (7-1)

$$\sigma_z = -K\frac{P}{z^2} = -K\frac{100}{z^2}$$

where by (7-2)

$$K = \frac{3z^5}{2\pi(r^2 + z^2)^{5/2}} = \frac{3z^5}{2\pi(25 + z^2)^{5/2}}$$

Then the desired variation of σ_z becomes a function of z; that is,

$$\sigma_z = -\frac{300z^3}{2\pi(25 + z^2)^{5/2}}$$

as shown in Fig. P-7.3.

7.6 A line load of constant intensity $q = 1000$ lb/ft is acting on the soil surface as shown in Fig. P-7.6. Find the general equation for the vertical stress σ_z in the subgrade at j. Given: $a = 30$ ft, $c = 0$, $d = 40$ ft.

Solution By (7-4)

$$\sigma_z = -\frac{3qz^3}{2\pi r_a^4}\left(\frac{d}{R_{ad}} - \frac{d^3}{3R_{ad}^3}\right)$$

where

$$r_a^4 = (30^2 + z^2)^2$$
$$R_{ad} = \sqrt{30^2 + 40^2 + z^2}$$

Then at j (any depth),

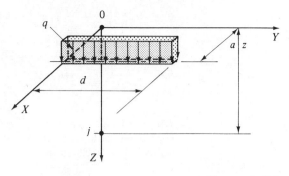

Figure P-7.6

Figure P-7.6

$$\sigma_z = -\frac{(3)(1000)z^3}{2\pi(30^2 + z^2)^2}\left[\frac{40}{(30^2 + 40^2 + z^2)^{1/2}} - \frac{40^3}{3(30^2 + 40^2 + z^2)^{3/2}}\right]$$

7.7 A rectangular load of constant intensity $q = 1000\ \text{lb/ft}^2$ is acting on the soil surface as shown in Fig. P-7.7. Find the general equation for the vertical stress in the subgrade at j. Given: $a = 0, b = 10\ \text{ft}, c = 0, d = 20\ \text{ft}$.

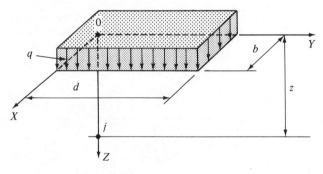

Figure P-7.7

Solution

By (7-5)

$$\sigma_z = -\frac{qbz}{2\pi d}\left(\frac{R_{bd}}{r_b^2} - \frac{z^2}{r_d^2 R_{bd}} + \frac{d}{bz}\tan^{-1}\frac{bd}{zR_{bd}}\right)$$

where

$$r_b^2 = \sqrt{b^2 + z^2} \qquad r_d^2 = \sqrt{d^2 + z^2}$$
$$R_{bd} = \sqrt{b^2 + d^2 + z^2}$$

Then at j (any depth).

$$\sigma_z = -\frac{q}{2\pi}\left[\frac{bdz^2}{r_b^2 r_a^2}\left(\frac{R_{bd}}{z} + \frac{z}{R_{bd}}\right) + \tan^{-1}\frac{bd}{zR_{bd}}\right]$$

$$= -\frac{1000}{2\pi}\left[\frac{(10)(20)z^2}{(10^2 + z^2)(20^2 + z^2)}\left(\frac{\sqrt{10^2 + 20^2 + z^2}}{z} + \frac{z}{\sqrt{10^2 + 20^2 + z^2}}\right)\right.$$

$$\left. + \tan^{-1}\frac{(10)(20)}{z\sqrt{10^2 + 20^2 + z^2}}\right]$$

7.8 A circular load of radius $a = 100$ in. and constant intensity $q = 50\,\text{lb/in.}^2$ is applied on the horizontal soil surface. Compute the central deflection $\delta_{jz}(z = 0)$. Given: $E_0 = 10,000\,\text{lb/in.}^2$

Solution By (4) of Table 7-2,

$$\delta_{jz} = \frac{3qa^2}{2E_0\sqrt{a^2 + z^2}}$$

At $z = 0$,

$$\delta_{jz}(z = 0) = \frac{(3)(50)(100)^2}{(2)(10,000)(100)} = 0.75\,\text{in.}$$

Influence Functions

7.9 A system of concentrated loads is applied on a horizontal soil surface as shown in Fig. P-7.9. Compute the vertical stress at j. Given: $P_1 = 10,000$ lb, $P_2 = 20,000$ lb, $P_3 = 30,000$ lb, $r_{01} = 2$ ft, $r_{02} = 4$ ft, $r_{03} = 6$ ft, $z_{0j} = 10$ ft.

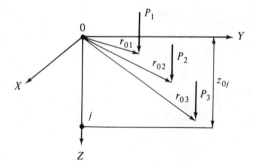

Figure P-7.9

Solution First, compute

$$\frac{r_{01}}{z_{0j}} = \frac{2}{10} = 0.2, \quad \frac{r_{02}}{z_{0j}} = \frac{4}{10} = 0.4, \quad \frac{r_{03}}{z_{0j}} = \frac{6}{10} = 0.6$$

Then, from Table 7-1, determine

$K_{ji} = 0.4329, \quad K_{j2} = 0.3294, \quad K_{j3} = 0.2214$

Finally, by superposition,

$$
\begin{aligned}
\sigma_{jz} &= \sigma_{j1z} + \sigma_{j2z} + \sigma_{j3z} \\
&= -\frac{K_{j1}P_1 + K_{j2}P_2 + K_{j3}P_3}{z_{0j}^2} \\
&= -\frac{(0.4329)(10,000) + (0.3294)(20,000) + (0.2214)(30,000)}{(10)^2} \\
&= -175.55 \text{ lb/in.}^2
\end{aligned}
$$

Shape Functions

7.10   A circular plate loaded with a total load of 60,000 lb is uniformly distributed over the surface at a rate of 100 lb/in². Calculate the vertical stress beneath the area at the following locations:

Depth z: 6, 12, 25, and 50 inches

Radial distance (from the center of the loaded area) r: 0, 3, 6, 12, 25, and 50 inches

The vertical stress σ_z at each location specified in the statement of this problem can be calculated by using (7-15):   *Solution*

$$\sigma_z = -p_z q \tag{a}$$

where p_z is the shape function of two dimensionless parameters m and n (Section 7-4) and q is the intensity of the uniform pressure on the loaded area. The following steps explain the use of equation (a) in the solution of this problem.

1. The radius of the circular loaded area must first be determined, as

$$a = \sqrt{\frac{P}{\pi q}} = \sqrt{\frac{(60,000)}{(\pi)(100)}} = 13.8 \text{ in.}$$

2. Since the coordinates of each point (r and z) are given, the position parameters $n = r/a$ and $m = z/a$ for each point are calculated.
3. The vertical stress coefficients p_z for each point are then determined by introducing the values of the position parameters m and n calculated in (2) into the chart in Table 7-5 and reading off the corresponding values of p_z.
4. Finally, the vertical stress σ_z is calculated by (7-15), as shown in Table P-7.10.

7.11   In Problem 7.10, draw the soil profile under the loaded area to an appropriate scale. Then show in a graphical form the distribution of the vertical stress σ_z out from the center of the loaded area on four horizontal planes at depths of 6, 12, 25, and 50 in. beneath the surface of the ground.

Solution

1. After drawing the soil profile to an appropriate scale, locate the 24 points designated in Table P-7.10.

TABLE P-7.10

POINT	COORDINATES (in.)		POSITION PARAMETERS		VERTICAL STRESS COEFFICIENT p_z	VERTICAL STRESS (lb/in²) σ_z
	r	z	$n = \dfrac{r}{a}$	$m = \dfrac{z}{a}$		
A_0	0	6	0	0.44	0.90	$-$ 90
A_1	3	6	0.22	0.44	0.89	$-$ 89
A_2	6	6	0.44	0.44	0.85	$-$ 85
A_3	12	6	0.87	0.44	0.60	$-$ 60
A_4	25	6	1.81	0.44	0.04	$-$ 4
A_5	50	6	3.62	0.44	Negligible	Negligible
B_0	0	12	0	0.87	0.70	$-$ 70
B_1	3	12	0.22	0.87	0.69	$-$ 69
B_2	6	12	0.44	0.87	0.66	$-$ 66
B_3	12	12	0.87	0.87	0.45	$-$ 45
B_4	25	12	1.81	0.87	0.05	$-$ 5
B_5	50	12	3.62	0.87	Negligible	Negligible
C_0	0	25	0	1.81	0.33	$-$ 33
C_1	3	25	0.22	1.81	0.32	$-$ 32
C_2	6	25	0.44	1.81	0.30	$-$ 30
C_3	12	25	0.87	1.81	0.24	$-$ 24
C_4	25	25	1.81	1.81	0.09	$-$ 9
C_5	50	25	3.62	1.81	Negligible	Negligible
D_0	0	50	0	3.62	0.10	$-$ 10
D_1	3	50	0.22	3.62	0.095	$-$ 9.5
D_2	6	50	0.44	3.62	0.09	$-$ 9
D_3	12	50	0.87	3.62	0.085	$-$ 8.5
D_4	25	50	1.81	3.62	0.070	$-$ 7
D_5	50	50	3.62	3.62	Negligible	Negligible

2. At each of these points, plot the vertical stress σ_z computed in Table P-7.10, using the same scale for plotting the stresses over all points.
3. Then connect all the stress ordinates, at each of the four levels, plotted in (2) to produce the required graphical presentation of the variation in the vertical stress σ_z.

These graphical presentations of the stress distribution at each of the four levels are shown in Fig. P-7.11.

7.12   On Fig. P-7.11, draw the isopressure surfaces (bulbs of pressure) representing the surfaces of equal stress, for every 10 percent of the surface vertical pressure [i.e., 90, 80, 70, ... etc. (lb/in.²) isopressure surfaces].

Solution   The isopressure surfaces are constructed by joining all the points under the surface of the ground that have the same vertical stress σ_z at the required level by a smooth curve. Here this step can be accomplished by using the calculations of the

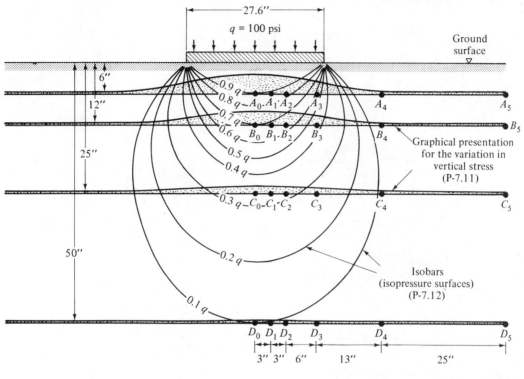

Figure P-7.11

vertical stress σ_z in Table P-7.10 and the graphical presentation of the distribution of vertical stress σ_z at different levels in Fig. P-7.11.

The isopressure surfaces at 10 percent intervals of pressure are shown in Fig. P-7.11.

7.13   The line load of intensity $q=10,000\,\text{lb/ft}$ is applied on the horizontal soil surface. Using the shape functions ψ_z, find the vertical stresses σ_z at points 1, 2, and 3. Given: $a=0, l=30\,\text{ft}, z=10\,\text{ft}$.

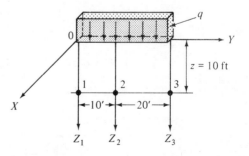

Figure P-7.13

Solution By (7-12)

$$\sigma_z = -q\psi_z$$

where

$$\psi_z = \psi_z(m, n), \quad m = \frac{a}{z}, \quad n = \frac{l}{z}$$

At 1,

$$m = 0, \quad n = \frac{30}{10} = 3$$

and from Table 7-3,

$$\psi_z(0, 3) = 0.318$$

which yields

$$\sigma_{1z} = -(1000)(0.318) = -318 \text{ lb/ft}^2$$

At 2,

$$m = 0, \quad n_{12} = \frac{l_{12}}{z} = \frac{10}{10} = 1, \quad n_{23} = \frac{l_{23}}{z} = \frac{20}{10} = z$$

and from Table 7-3,

$$\psi_{12z}(0, 1) = 0.049 \qquad \psi_{23z}(0, 2) = 0.315$$

which yields

$$\sigma_{2z} = -(1000)(0.049 + 0.315) = -363 \text{ lb/ft}^2$$

At 3,

$$m = 0 \quad n = \frac{30}{10} = 3$$

and

$$\sigma_{3z} = \sigma_{1z} = -318 \text{ lb/ft}^2$$

7.14 Consider the line load of Problem 7.13, placed as shown in Fig. P-7.14. Using the shape functions ψ_z, find the vertical stress at *j*. Given: $a = 0, c = 10$ ft, $d = 40$ ft, $z = 10$ ft.

Solution By (7-12)

$$\sigma_z = -\frac{q[\psi_z(0, 40/10) - \psi_z(0, 10/10)]}{z}$$

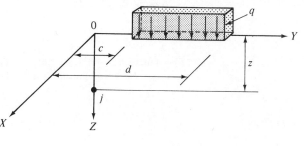

Figure P-7.14

where

$$\psi_z(0, 4) = 0.319 \qquad \psi_z(0, 1) = 0.280$$

Thus

$$\sigma_z = -1000(0.319 - 0.280) = -39 \, \text{lb/ft}^2$$

7.15   A rectangular area 100 ft long and 50 ft wide is loaded with a uniform pressure of 400 lb/in.² and rests over a semi-infinite mass of rock (Fig. P-7.15). Compute the vertical stresses σ_z along the centerline of the loaded area and at depth intervals of 5 ft and up to a total depth of 70 ft.

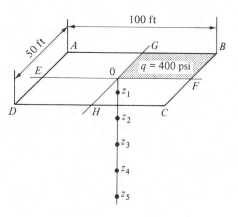

Figure P-7.15

Solution

1. The total area *ABCD* is divided into four separate and equal-sized areas of 50 ft by 25 ft each. The four areas are abutting with their common corner at the center of the loaded area *ABCD*.
2. The vertical stress σ_z at the corner of each of the four areas is computed by

using Eq. (7-13); that is,

$$\sigma_z = -q\phi_z(m, n) \tag{a}$$

where $q = 400 \text{ lb/in.}^2$ and $\phi_z(m, n)$ is the shape function of the dimensionless parameters m and n. In this case,

$$m = \frac{v}{z} = \frac{25}{z} \quad \text{and} \quad n = \frac{u}{z} = \frac{50}{z} \tag{b}$$

For each depth z, the values of m and n are computed and are then introduced in the chart of Table 7-4, where the values of ϕ_z are determined. Substituting these values of q and ϕ_z in equation (a), the value of σ_z at each depth is determined.

3. The total stress at the common corner of these four areas is determined by summing up the stresses calculated from each area separately, as explained in step (2). However, since the four areas in this problem are equal, only one is considered and the resulting stress is multiplied by 4.

Details of computations are shown in Table P-7.15.

TABLE P-7.15

Point	Depth z (ft)	Length of Loaded Area u (ft)	Width of Loaded Area v (ft)	$m = \frac{v}{z}$	$n = \frac{u}{z}$	Influence Value ϕ_z	Vertical Stress $\sigma'_z = -\phi q$ (lb/in.²)	Total Vertical Stress $(4)(\sigma'_z)$ (lb/in.²)
z_1	5	50	25	10	5	0.247	− 98.8	− 395.2
z_2	10	50	25	5	2.5	0.244	− 97.6	− 390.4
z_3	15	50	25	3.33	1.67	0.232	− 92.8	− 371.2
z_4	20	50	25	2.50	1.25	0.217	− 86.8	− 347.2
z_5	25	50	25	2.0	1.00	0.199	− 79.6	− 318.4
z_6	30	50	25	1.67	0.83	0.180	− 72.0	− 288.0
z_7	35	50	25	1.43	0.71	0.163	− 65.2	− 260.8
z_8	40	50	25	1.25	0.63	0.148	− 59.2	− 236.8
z_9	45	50	25	1.11	0.56	0.132	− 52.8	− 211.2
z_{10}	50	50	25	1.00	0.50	0.121	− 48.4	− 193.6
z_{11}	55	50	25	0.99	0.46	0.112	− 44.8	− 179.2
z_{12}	60	50	25	0.83	0.42	0.099	− 39.6	− 158.2
z_{13}	65	50	25	0.77	0.38	0.087	− 34.8	− 139.2
z_{14}	70	50	25	0.71	0.36	0.079	− 31.6	− 126.4

7.16 A rectangular area 50 ft long and 20 ft wide is loaded with a uniform pressure of $1.0t/\text{ft}^2$. The loaded area is resting over a semi-infinite homogeneous soil mass. Calculate the vertical normal stress σ_z below the point G located outside the loaded area (Fig. P-7.16), at depths of 20 ft and 50 ft below the ground surface.

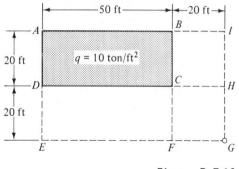

Figure P-7.16

Solution

1. Assume that the entire area *AEGI* is loaded with a uniform pressure of $q = 1.0t/ft^2$ and let $\sigma_{1z}, \sigma_{2z}, \sigma_{3z}$, and σ_{4z} be the vertical stresses beneath point *G*, at depths $z = 20$ ft due to loaded areas *AEGI, DEGH, BFGI*, and *CFGH* respectively.
2. Since point *G* is now a common corner to the four separate areas in step (1), Fadum's chart (Table 7-4) can be used for calculating the vertical stress $\sigma_{1z}, \ldots, \sigma_{4z}$, given by (7-13) as

$$\sigma_z = -\phi_z(m, n)q \qquad (a)$$

in which *m* and *n* are the shape parameters (defined in Table 7-4) for each of the four areas. After *m* and *n* are determined, their values are introduced in the chart (Table 7-4) and from which $\phi(m, n)$ are determined for each area. Equation (a) in terms of $q = 1.0t/ft^2$ and of computed $\phi(m, n)$, yields σ_z.
3. The total vertical stress σ_z beneath *G* is

$$\sigma_z = (\sigma_{1z}) + (-\sigma_{2z}) + (-\sigma_{3z}) + (\sigma_{4z})$$

4. The same steps are repeated for calculating the stresses at the 50 ft depth.

Table P-7.16 shows these calculations.

7.17 For the same loaded area and load distribution as in Problem 7.16, calculate the total vertical stress σ_z at a depth of 50 ft beneath the point *G*, using Newmark's influence chart (Fig. 7.9). Compare the answer with that obtained in Problem 7.16.

In order to solve this problem using Newmark's influence chart (Fig. 7-9), the   *Solution* plan of the loaded area *ABCD* is drawn to such a scale that the length *OQ*, shown at the top of the chart in Fig. 7-9, represents a depth of 50 ft. The plan of the loaded area, drawn on a transparent sheet, is then placed on the top of the chart in the figure in such a way that the point beneath which the stress calculation is desired (in this case, point *G*) will be over the center of the chart.

The number of influence areas (including fractions) covered by the plan is counted and introduced in (7-20).

$$\sigma_z = (N)(q)(i)$$

in which N is 32.5 areas; q is $1.0\ t/\text{ft}^2$; and i is 0.001. Substituting these quantities in (7-20) yields

$$\sigma_z = (32.5)(1)(0.001) = -0.0325\ t/\text{ft}^2$$

This value of σ_z is very close to the one obtained in Problem 7.16.

TABLE P-7.16

Area	Length u (ft)	Breadth v (ft)	Depth z (ft)	$m=\dfrac{u}{z}$	$n=\dfrac{v}{z}$	In-fluence Value ϕ	Vertical Stress $\sigma_z = -(\phi)(q)$	Depth z (ft)	$m=\dfrac{u}{z}$	$n=\dfrac{v}{z}$	In-fluence Value ϕ	Vertical Stress $\sigma_z = -(\phi)(q)$
				FOR DEPTH z OF 20 ft					FOR DEPTH z OF 50 ft			
$+AEGI$	70	40	20	3.5	2.00	+0.238	−0.238	50	1.4	0.8	+0.1720	−0.1720
$+CFGH$	20	20	20	1.00	1.00	+0.175	−0.175	50	0.4	0.4	+0.0610	−0.0610
$-DEGH$	70	20	20	3.5	1.00	−0.203	+0.203	50	1.4	0.4	−0.1075	+0.1075
$-BFGI$	40	20	20	2.00	1.00	−0.199	+0.199	50	0.8	0.4	−0.0930	+0.0930
			Total stress $\sigma_z = -0.011\ t/\text{ft}^2$						Total stress $\sigma_z = -0.0325\ t/\text{ft}^2$			

SUPPLEMENTARY PROBLEMS

Single Concentrated Load

7.18   Compute the vertical stress at a depth of 15 ft and along the line of action of a concentrated load of 100 tons acting vertically on the surface of a homogeneous, isotropic, and semi-infinite mass of soil.

Answer

$$\sigma_z = 0.21\ t/\text{ft}^2$$

Line Load of Constant Intensity

7.19   Evaluate the integral of (7-4), given as

$$\int \frac{dy}{(a^2 + y^2 + z^2)^{5/2}} = \int \frac{dy}{(r_a^2 + y^2)^{5/2}}$$

where

$$r_a^2 = a^2 + z^2$$

Hint: Integrate by parts as

$$\frac{y}{3r_a^2(r_a^2 + y^2)^{3/2}} + \frac{3}{2r_a^2}\int \frac{dy}{(r_a^2 + y^2)^{3/2}}$$

Circular Load of Constant Intensity

7.20 Two circular rigid slabs are resting on the surface of a semi-infinite, homo-geneous, and isotropic soil mass, spaced 37.5 in. center to center, as shown in Fig. P-7.20. Each slab is considered weightless and carries a 50-kip load with a uniform pressure of 100 lb/in². Calculate the vertical stresses σ_z beneath the center of one area at depths of 4, 8, 16, 25, 36, 48, 60, and 72 in.

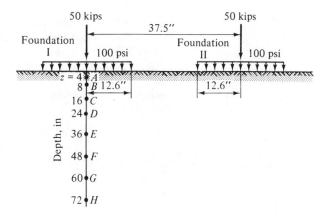

Figure P-7.20

Referring to Fig. P-7.20, the vertical stresses σ_z in lb/in² at points A, B, C, D, E, F, *Answer* G, and H are 0, 0.25, 1.00, 2.30, 3.10, 3.20, 3.00, and 2.60 respectively.

Shape Functions

√**7.21** A rectangular rigid slab 12 ft long and 9 ft wide carries a uniformly distri-buted pressure of $2.0\ t/ft^2$ and rests on a semi-infinite, homogeneous soil mass. Com-pute the intensity of vertical pressure σ_z at a depth of 10 ft beneath point O, situated as shown in Fig. P-7.21.

The student may try several approaches to the solution of this problem and then *Answer* compare the results as follows:

1. The area can be divided up into 12 smaller areas, 3 ft square each. The load on each area will then be considered as a concentrated load of 18 tons each, located at the center of each individual area. Then (7-1) is used to determine the vertical stress σ_z beneath point O due to each load and the results are added to determine the total stress due to all 12 loads.

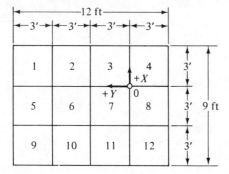

Figure P-7.21

2. Fadum's influence chart (Table 7-4) can be used as in Problem 7.15 by dividing the area into four areas with their common corner at point O. The stress at point O due to the load on each area is then calculated and the results are added to obtain the total vertical stress at O due to the load covering the entire area.
3. Newmark's influence chart (Fig. 7-9) can also be used to solve this problem, as in Problem 7.17.

All approaches will lead to the approximate value, $\sigma_z = -0.627 \, t/\text{ft}^2$.

strength analysis

8

8.1 TRIAXIAL STATE

Stress Tensor

The state of stress at a given point in the soil system is obtained by removing from the subgrade a small rectangular parallelepiped, the edges of which (dx, dy, dz) are parallel to the respective coordinate axes (Fig. 8-1). The stress components acting on the respective faces of this parallelepiped form the array given in Table 8-1, called the *stress tensor* (stress dyadic). For uniformity (of writing), all stress components in this chapter have two subscripts and one superscript. The first subscript indicates the direction normal to the face of the parallelepiped, the second one designates the direction of the stress component, and the superscript identifies the coordinate system.

Stress Transformations

As different sections are passed through the same point, different stresses occur on the elemental areas of different orientation. If the stress tensors T^m and T^n at the same point are related to the systems

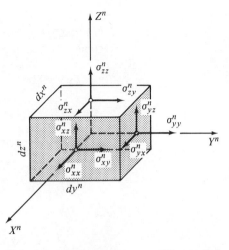

Stress tensor in n-system   **Figure 8-1**

TABLE 8-1 *Stress Tensor*

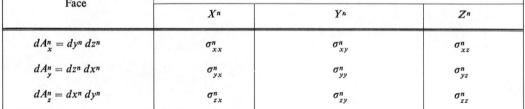

Face	STRESS DIRECTION		
	X^n	Y^n	Z^n
$dA^n_x = dy^n\,dz^n$	σ^n_{xx}	σ^n_{xy}	σ^n_{xz}
$dA^n_y = dz^n\,dx^n$	σ^n_{yx}	σ^n_{yy}	σ^n_{yz}
$dA^n_z = dx^n\,dy^n$	σ^n_{zx}	σ^n_{zy}	σ^n_{zz}

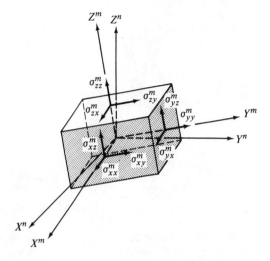

Figure 8-2   *Stress tensor in m-system*

X^m, Y^m, Z^m (Fig. 8-2) and X^n, Y^n, Z^n (Fig. 8-1), respectively, their relationships become

$$T^m = \Pi^{mn} T^n \Pi^{nm} \qquad T^n = \Pi^{nm} T^m \Pi^{mn} \tag{8-1, 2}$$

where

$$T^m = \begin{bmatrix} \sigma^m_{xx} & \sigma^m_{xy} & \sigma^m_{xz} \\ \sigma^m_{yx} & \sigma^m_{yy} & \sigma^m_{yz} \\ \sigma^m_{zx} & \sigma^m_{zy} & \sigma^m_{zz} \end{bmatrix} \qquad T^n = \begin{bmatrix} \sigma^n_{xx} & \sigma^n_{xy} & \sigma^n_{xz} \\ \sigma^n_{yx} & \sigma^n_{yy} & \sigma^n_{yz} \\ \sigma^n_{zx} & \sigma^n_{zy} & \sigma^n_{zz} \end{bmatrix} \tag{8-3, 4}$$

and

$$\Pi^{nm} = \begin{bmatrix} \alpha_x & \alpha_y & \alpha_z \\ \beta_x & \beta_y & \beta_z \\ \gamma_x & \gamma_y & \gamma_z \end{bmatrix} \qquad \Pi^{mn} = \begin{bmatrix} \alpha_x & \beta_x & \gamma_x \\ \alpha_y & \beta_y & \gamma_y \\ \alpha_z & \beta_z & \gamma_z \end{bmatrix} \tag{8-5, 6}$$

The coefficients of the angular transformation matrices Π^{nm}, Π^{mn} are the direction cosines, given below as

$$\alpha_x = \cos(X^n, X^m), \quad \alpha_y = \cos(X^n, Y^m), \quad \alpha_z = \cos(X^n, Z^m)$$
$$\beta_x = \cos(Y^n, X^m), \quad \beta_y = \cos(Y^n, Y^m), \quad \beta_z = \cos(Y^n, Z^m) \tag{8-7}$$
$$\gamma_x = \cos(Z^n, X^m), \quad \gamma_y = \cos(Z^n, Y^m), \quad \gamma_z = \cos(Z^n, Z^m)$$

The angular transformation matrices are orthogonal; that is,

$$\Pi^{nm} = \Pi^{mn)T} = \Pi^{mn)-1} \qquad \Pi^{mn} = \Pi^{nm)T} = \Pi^{nm)-1} \tag{8-8}$$

where Π^T and Π^{-1} are the transpose and the inverse of Π respectively (Problem 8.1).

From the conditions of moment equilibrium of the volume element (Fig. 8-1),

$$\sigma^n_{xy} = \sigma^n_{yx}, \quad \sigma^n_{yz} = \sigma^n_{zy}, \quad \sigma^n_{zx} = \sigma^n_{xz} \tag{8-9}$$

and T^n is a symmetrical tensor, six components of which completely determine the state of stress at a point. Since the angular transformation matrices are orthogonal, the congruent transformations in (8-1) and (8-2) preserve the symmetry and T^m is again a symmetrical tensor.

Principal Stresses

For any general state of stress through a given point in a soil system, there exist three mutually perpendicular planes on which the shearing stress vanishes. Such planes are called the *principal planes*, their normals are the *principal axes* (X^p, Y^p, Z^p), and the stresses acting along these axes are the *principal stresses* $(\sigma^p_1, \sigma^p_2, \sigma^p_3)$.

The principal stresses are the extremes of the normal stress at that point and their values $\sigma^p_1, \sigma^p_2, \sigma^p_3$ are the roots of the determinant equation given below.

$$\begin{vmatrix} \sigma^n_{xx} - \sigma^p_{1,2,3} & \sigma^n_{xy} & \sigma^n_{xz} \\ \sigma^n_{yx} & \sigma^n_{yy} - \sigma^p_{1,2,3} & \sigma^n_{yz} \\ \sigma^n_{zx} & \sigma^n_{zy} & \sigma^n_{zz} - \sigma^p_{1,2,3} \end{vmatrix}$$
$$= (\sigma^p_{1,2,3})^3 - J_1(\sigma^p_{1,2,3})^2 + J_2(\sigma^p_{1,2,3}) - J_3 = 0 \tag{8-10}$$

where

$$J_1 = \sigma^n_{xx} + \sigma^n_{yy} + \sigma^n_{zz}$$
$$J_2 = \begin{vmatrix} \sigma^n_{xx} & \sigma^n_{xy} \\ \sigma^n_{yx} & \sigma^n_{yy} \end{vmatrix} + \begin{vmatrix} \sigma^n_{yy} & \sigma^n_{yz} \\ \sigma^n_{zy} & \sigma^n_{zz} \end{vmatrix} + \begin{vmatrix} \sigma^n_{zz} & \sigma^n_{zx} \\ \sigma^n_{xz} & \sigma^n_{xx} \end{vmatrix} \tag{8-11}$$
$$J_3 = \begin{vmatrix} \sigma^n_{xx} & \sigma^n_{xy} & \sigma^n_{xz} \\ \sigma^n_{yx} & \sigma^n_{yy} & \sigma^n_{yz} \\ \sigma^n_{zx} & \sigma^n_{zy} & \sigma^n_{zz} \end{vmatrix}$$

are the invariants of the stress tensor; they are not affected by the transformation of coordinate axes (Problem 8.2).

Extreme Shearing Stresses

The extreme shearing stresses occur on the planes containing one principal axis and bisecting the angle of the two remaining principal axes; they are

$$\sigma^s_{1,23} = \pm\frac{\sigma^p_2 - \sigma^p_3}{2}, \quad \sigma^s_{2,31} = \pm\frac{\sigma^p_3 - \sigma^p_1}{2}, \quad \sigma^s_{3,12} = \pm\frac{\sigma^p_1 - \sigma^p_2}{2} \tag{8-12}$$

where $\sigma^p_{1,2,3}$ are the principal stresses given by (8-10) and s is the extreme shear system.

Independence

All relationships of this section have been exclusively derived from the conditions of static equilibrium and consequently apply regardless of the material involved. The same is valid in the biaxial relationships derived in the subsequent section.

8.2 BIAXIAL STATE

Stress Tensor

The results of Section 8.1 indicate that the extreme shearing stress given by (8-11) is determined by the *biaxial state* if the two respective principal stresses are known. Furthermore, the results of Section 7.2 reveal that the loaded soil system is (in general) in a state of compression; that is,

$$\sigma_1 < 0, \quad \sigma_2 < 0, \quad \sigma_3 < 0, \quad |\sigma_1| > |\sigma_2| > |\sigma_3|.$$

For reasons of simplicity, in engineering soil mechanics the customary sign convention of continuum mechanics was changed and the stresses of Fig. 8-3 were designated as positive (this is obviously an inconsistent but convenient sign convention).

The stresses in the *m*-system, derived by reduced (8-1) in terms of the trigonometric equivalents of (8-7) and the sign convention of Fig. 8-3, are

$$\sigma^m_{xx} = \frac{\sigma^p_1 + \sigma^p_3}{2} - \frac{\sigma^p_1 - \sigma^p_3}{2}\cos 2\omega \tag{8-13}$$

$$\sigma^m_{xz} = \sigma^m_{zx} = \frac{\sigma^p_1 - \sigma^p_3}{2}\sin 2\omega \tag{8-14}$$

$$\sigma^m_{zz} = \frac{\sigma^p_1 + \sigma^p_3}{2} + \frac{\sigma^p_1 - \sigma^p_3}{2}\cos 2\omega \tag{8-15}$$

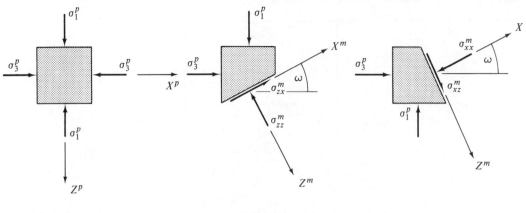

(a) Principal stress in p-system   (b) Normal and shearing stresses in m-system

Biaxial state of stress   **Figure 8-3**

where σ_1^p, σ_3^p are the maximum and minimum principal stresses, respectively, and ω is the angle of X^p, X^m.

Extreme Shearing Stresses

The maximum shearing stress occurs when $\sin 2\omega = 1$, or $\omega = \pi/4, 5\pi/4$, and is

$$\sigma_{xz,\max}^s = \frac{\sigma_1^p - \sigma_3^p}{2} \tag{8-16}$$

The normal stress on the plane of maximum shear ($\cos 2\omega = 0$) is

$$\sigma_{xx}^s = \sigma_{zz}^s = \frac{\sigma_1^p + \sigma_3^p}{2} \tag{8-17}$$

which indicates that, for $\omega = \pi/4, 5\pi/4$, (8-13) and (8-15) yield the same result (Table 8-2).

Mohr's Circle

The graphical representation of the stress variation at a point in the soil system is given by *O. Mohr's Circle* (1882). The equation of this circle, given as

$$\left(\sigma_{zz}^m - \frac{\sigma_1^p + \sigma_3^p}{2}\right)^2 + (\sigma_{zx}^m)^2 = \left(\frac{\sigma_1^p - \sigma_3^p}{2}\right)^2 \tag{8-18}$$

is the sum of the squares of (8-14) and (8-15), with the terms containing the trigonometric function left on the right side of each equation.

TABLE 8-2   *Mohr's Circle Construction*

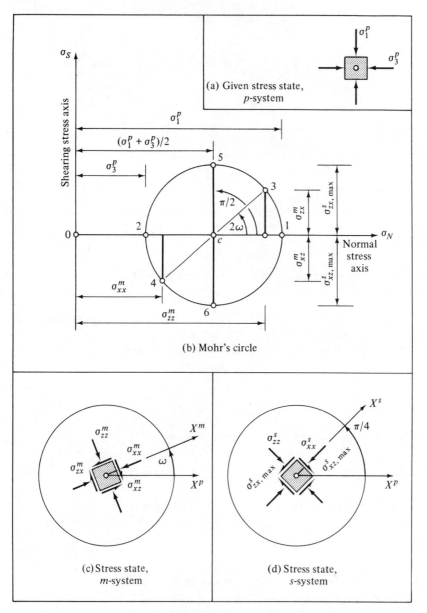

(a) Given stress state,
p-system

(b) Mohr's circle

(c) Stress state,
m-system

(d) Stress state,
s-system

The geometry of the circle is related to two orthogonal axes σ_N and σ_S. In this system, the horizontal coordinates represent σ_{xx}^m, σ_{zz}^m, the vertical coordinates represent σ_{xz}^m, σ_{zx}^m, the position of the center of the circle is given by $[(\sigma_1^p + \sigma_3^p)/2, 0]$, and its radius is $(\sigma_1^p - \sigma_3^p)/2$.

The construction of the circle is shown in Table 8-2, and its application to the solution of a biaxial state of stress is given in Problem 8.4.

8.3 *STRENGTH THEORIES AND LABORATORY TESTS*

General

The location of planes and magnitudes of the extreme stresses precedes the ultimate aim of the analysis—the determination of strength of the subgrade at a given point.

By definition, the *strength of a soil system* is its ultimate capacity to resist (carry, support, bear) the stresses developed by causes (applied loads, over-burden of overlaying soil, hydrostatic pressure, etc.). This property is not easy to determine because of the complex nature of the material and the discrepancies between the properties of the test specimen and those of the soil in the system.

Strength Theories

The science of strength of materials developed a number of theories of strength (of failure), none of which explains the strength of all materials; consequently, none is completely satisfactory.

One of these theories, *Mohr's Theory of Rupture* (1900), however, gives results verified by experiments with soils and satisfies the needs of engineering soil mechanics as an indicator of soil strength. According to Mohr, the failure within the soil system is not caused by the extreme stress of one type (normal stress or shearing stress) but by a critical combination of both.

The line connecting the critical combinations of the normal stress and shearing stress is known as *Mohr's Envelope of Rupture*, and the failure will occur when, for the attained normal stress, the associated shearing stress penetrates the envelope (Fig. 8-4). Although this envelope is a curve, it can be approximated by a straight line (for practical purposes).

From the geometry of Fig. 8-4,

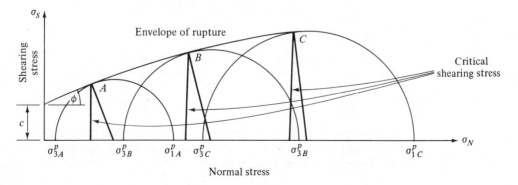

Mohr's envelope of rupture   **Figure 8-4**

$$\sigma_1^p = 2c \tan\left(\frac{\pi}{4} + \frac{\phi}{2}\right) + \sigma_3^p \tan^2\left(\frac{\pi}{4} + \frac{\phi}{2}\right) \qquad (8\text{-}19)$$

where c, ϕ are the parameters of the envelope (determined by experiments) and σ_1^p, σ_3^p are the principal stresses in the p-system.

The general application of (8-19) is shown in Problem 8.5; special forms of this equation for cohesionless and cohesive soils are introduced in Sections 8.4 and 8.5 respectively.

Strength Laboratory Tests

Several soil strength tests have been devised throughout the years; the principal tests in this category are

1. The *direct shear test* (Problems 8.6, 8.7, 8.8) is the oldest shear test. It is still extensively used. The resulting strength of the sample is usually higher than the actual strength of the natural soil system.
2. The *ring shear test* (Problem 8.9) allows a better control of water content in small samples, but it has the deficiencies of the direct shear test.
3. The *triaxial shear test* (Problem 8.10) is the most reliable shear test. The special quality of this test is the uniformity of stress distribution and the freedom of the sample to fail along the weakest plane.

If σ_3 in the last test is zero, the designation of *unconfined compression test* is used.

8.4   STRENGTH OF COHESIONLESS SOILS

Dry Cohesionless Soils

The study of soil strength carried on in the preceding sections *disregarded* the structure of soil, its moisture content, and the properties of mineral grains forming the solid phase of the system. In this section, the influence of these properties is considered.

In cohesionless soils composed of rigid particles (gravels, sand, and/or silts) that are not exceptionally loose, the strength equation is

$$s = \bar{\sigma}_{zx} = \bar{\sigma}_{zz} \tan\phi = p \tan\phi \qquad (8\text{-}20)$$

where $s = \bar{\sigma}_{zx}$, $p = \bar{\sigma}_{zz}$ are the critical stresses and ϕ is the slope of the rupture line, given in this case as a straight line passing through the origin (Fig. 8-5).

Angle of Internal Friction

The angle ϕ is called the *angle of internal friction* and its typical values are given in Table 8-3 for $\sigma_{zz} = 5\ \mathrm{kg/cm^2}$. For each additional increment of $5\ \mathrm{kg/cm^2}$, the values of ϕ decrease approximately by 1 degree. The reason is attributed

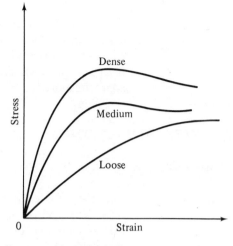

Stress-strain diagrams, cohesionless soils   **Figure 8-5**

Typical Values of ϕ in Degrees   **TABLE 8-3**

Material	Loose	Dense
Sand and gravel	33–36	47–50
Well-graded sand	30–33	40–47
Uniform fine-medium sand	26–30	32–38
Silty sand	26–30	30–35

mainly to the percentage of crushed grains as the pressure increases (Problem 8.11).

Wet Cohesionless Soils

In saturated cohesionless soils, the normal stress at failure consists of the inter-granular stress p_g and the hydrostatic stress u_w. Consequently, (8-20) becomes

$$s = \bar{\sigma}_{zx} = \underbrace{(p_g - u_w)}_{\bar{\sigma}_{zz}} \tan \phi = p \tan \phi \tag{8-21}$$

where ϕ is the same value as in dry soil (Problems 8.12, 8.13).

The hydrostatic effect may lead to the metastable conditions discussed at the end of Chapter 5. In such cases, the strength ceases to be a factor in foundation design.

Stress-Strain Diagram

The stress-strain relationships of soils bear some similarity to those of other engineering materials. The graphical representation of this relationship is the stress-strain curve (Fig. 8-5). The slope of this curve is the modulus of the soil. Since it varies along the curve, its values (given by Table 6-1) are only crude approximations.

8.5 *STRENGTH OF COHESIVE SOILS*

Moisture Dissipation

The study of strength of clays is far more involved than that of cohesionless soils. The load-bearing agent of clays is first the soil water, which carries the load as a cushion, allowing only a slow rate of volume change (because of low permeability).

For this reason, the shear strength in clays is defined in terms of *moisture dissipation* and *degree consolidation*. Three typical tests are available for the strength test of clays :

1. the *consolidated-drained shear test* (Problem 8.14).
2. the *consolidated-undrained shear test* (Problem 8.15).
3. the *unconsolidated-undrained shear test* (Problem 8.16).

These laboratory tests serve as simulators of natural situations in the soil systems and their results at best are approximations.

Sensitivity

The sensitivity of clay is a peculiar characteristic defined as the ratio of the strength of undisturbed to the remolded specimen of the same clay, called the degree of sensitivity. For clays, this ratio ranges from 2 to 4 (low sensitivity), 4 to 8 (medium sensitivity), and over 8 (extrasensitive clays).

Angle of Internal Friction

The experimental results of drained strength tests show that for normally loaded clays of low and medium sensitivity, $c = 0$ in the line of rupture, and the critical shear is

$$s = \bar{\sigma}_{zx} = \bar{\sigma}_{zz} \tan \phi = p \tan \phi \tag{8-22}$$

where $p = \bar{\sigma}_{zz}$ is the critical normal stress and ϕ is the angle of internal friction related to the plasticity index as shown below (Problem 8.17).

Plasticity index	0	20	40	80	100
ϕ in degrees	40–35	35–50	30–26	26–24	24–20

In cases of overconsolidated clays, (8-22) changes to

$$s = \bar{\sigma}_{zx} = c_1 + \bar{\sigma}_{zz} \tan \phi_1 = c_1 + p \tan \phi_1 \qquad (8\text{-}23)$$

where c_1, ϕ_1 are empirical constants (Problems 8.18, 8.19). Since partially saturated clays become saturated eventually, (8-23) is of a limited value, and in engineering calculations $c_1 \cong 0$ and $\phi_1 \cong \phi$.
   Finally, a special case arises when $\phi = 0$, but $c \neq 0$. This case, known as the *A. W. Skempton Condition* (1941), leads to

$$s = \bar{\sigma}_{zx} = c = \frac{\sigma_1^p - \sigma_3^p}{2} \qquad (8\text{-}24)$$

where σ_1^p, σ_3^p are the principal stresses and the rupture plane is at 45 degrees with the Z axis regardless of their magnitude. The importance of this condition in the soil system and the laboratory test equivalent are discussed in Problems 8.20 through 8.27.

Stress-Strain Diagram

The stress-strain curves of clays resemble those of some metals, and, for the same clay, the undisturbed and remolded samples present two entirely different

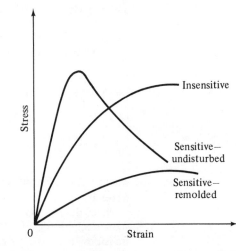

Stress-strain diagrams, cohesive soils   **Figure 8-6**

curves (Fig. 8-6). Since the slopes of these curves are also different, the modulus of subgrade must be taken as the slope of the undisturbed sample curve.

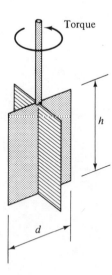

Torque

h

d

Figure 8-7
Vane shear test

In addition to the laboratory tests discussed earlier, it is often desirable to estimate the shear strength of a subgrade or a foundation soil in its undisturbed condition in the field. For this purpose, an *in situ* shear strength test known as the *vane test* is used; it can also be used as a laboratory test. A vane, similar to that illustrated in Fig. 8-7, is driven into the soil at the bottom of a prepared borehole. A torque is then applied at the surface and measured by a special gage. A cylinder of soil will resist the torque until it fails. The governing relationship is

$$\text{Torque } T = c\pi \left(\frac{d^2 h}{2} + \frac{d^3}{2} \right) \tag{8-25}$$

in which c is the cohesion of soil, and d, h are dimensions of the vane as illustrated in Fig. 8-7.

The use of this relationship is demonstrated in Problem 8.28. This test is found to agree with triaxial tests at depths between 50 and 100 ft. A smaller-sized vane is used when the test is conducted in the laboratory.

SOLVED PROBLEMS

Triaxial State

8.1 Derive the coefficients of the angular transformation matrices Π^{mn}, Π^{nm} (8-5, 8-6) and show their relationships (8-7).

Solution The transformation relations between the vector components in the X^m, Y^m, Z^m axes and their counterparts in the X^n, Y^n, Z^n axes (Fig. 8-2) and vice versa are given by the angular transformation matrices as

$$\underbrace{\begin{bmatrix} F_x^n \\ F_y^n \\ F_z^n \end{bmatrix}}_{V^n} = \underbrace{\begin{bmatrix} \alpha_x & \alpha_y & \alpha_z \\ \beta_x & \beta_y & \beta_z \\ \gamma_x & \gamma_y & \gamma_z \end{bmatrix}}_{\Pi^{nm}} \underbrace{\begin{bmatrix} F_x^m \\ F_y^m \\ F_z^m \end{bmatrix}}_{V^m} \qquad \underbrace{\begin{bmatrix} F_x^m \\ F_y^m \\ F_z^m \end{bmatrix}}_{V^m} = \underbrace{\begin{bmatrix} \alpha_x & \beta_x & \gamma_x \\ \alpha_y & \beta_y & \gamma_y \\ \alpha_z & \beta_z & \gamma_z \end{bmatrix}}_{\Pi^{mn}} \underbrace{\begin{bmatrix} F_x^n \\ F_y^n \\ F_z^n \end{bmatrix}}_{V^n} \qquad \text{(a, b)}$$

where F_x^m, F_y^m, F_z^m and F_x^n, F_y^n, F_z^n are components of a vector V (any vector) in the respective system.

Since the position of X^m, Y^m, Z^m can always be attained by rotating X^n, Y^n, Z^n successively through $\theta_x, \theta_y, \theta_z$, as shown in Fig. P-8.1, the coefficients of the angular transformation matrices must be a result of a triple matrix product, each of which

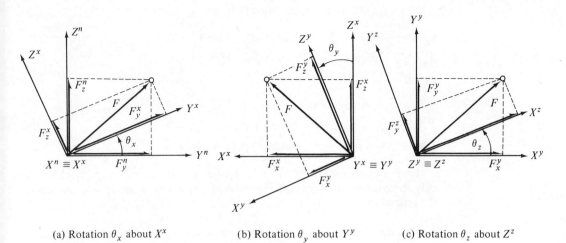

(a) Rotation θ_x about X^x (b) Rotation θ_y about Y^y (c) Rotation θ_z about Z^z

Figure P-8.1

corresponds to one rotation about one axis; that is,

$$
\Pi^{nm} = \begin{bmatrix} 1 & 0 & 0 \\ 0 & \cos\theta_x & -\sin\theta_x \\ 0 & \sin\theta_x & \cos\theta_x \end{bmatrix} \begin{bmatrix} \cos\theta_y & 0 & \sin\theta_y \\ 0 & 1 & 0 \\ -\sin\theta_y & 0 & \cos\theta_y \end{bmatrix} \begin{bmatrix} \cos\theta_z & -\sin\theta_z & 0 \\ \sin\theta_z & \cos\theta_z & 0 \\ 0 & 0 & 1 \end{bmatrix}
$$

$$
= \begin{bmatrix} \alpha_x & \alpha_y & \alpha_z \\ \beta_x & \beta_y & \beta_z \\ \gamma_x & \gamma_y & \gamma_z \end{bmatrix} \tag{c}
$$

$$
\Pi^{mn} = \begin{bmatrix} \cos\theta_z & \sin\theta_z & 0 \\ -\sin\theta_z & \cos\theta_z & 0 \\ 0 & 0 & 1 \end{bmatrix} \begin{bmatrix} \cos\theta_y & 0 & -\sin\theta_y \\ 0 & 1 & 0 \\ \sin\theta_y & 0 & \cos\theta_y \end{bmatrix} \begin{bmatrix} 1 & 0 & 0 \\ 0 & \cos\theta_x & \sin\theta_x \\ 0 & -\sin\theta_x & \cos\theta_x \end{bmatrix}
$$

$$
= \begin{bmatrix} \alpha_x & \beta_x & \gamma_x \\ \alpha_y & \beta_y & \gamma_y \\ \alpha_z & \beta_z & \gamma_z \end{bmatrix} \tag{d}
$$

Equations (c) and (d) show that Π^{nm}, Π^{mn} are orthogonal and satisfy (8-8).* In their derivations,

$$
F_x^z = F_x^m, \quad F_y^z = F_y^m, \quad F_z^z = F_z^m \tag{e}
$$

*Jan J. Tuma, *Engineering Mathematics Handbook*, McGraw-Hill, New York, 1970, p. 57.

From the orthogonality condition (8-8),

$$\alpha_x^2 + \alpha_y^2 + \alpha_z^2 = 1 \qquad\qquad \alpha_x^2 + \beta_x^2 + \gamma_x^2 = 1$$
$$\beta_x^2 + \beta_y^2 + \beta_z^2 = 1 \qquad\qquad \alpha_y^2 + \beta_y^2 + \gamma_y^2 = 1 \qquad\qquad \text{(f, g)}$$
$$\gamma_x^2 + \gamma_y^2 + \gamma_z^2 = 1 \qquad\qquad \alpha_z^2 + \beta_z^2 + \gamma_z^2 = 1$$

$$\alpha_x\beta_x + \alpha_y\beta_y + \alpha_z\beta_z = 0 \qquad \alpha_x\alpha_y + \beta_x\beta_y + \gamma_x\gamma_y = 0$$
$$\beta_x\gamma_x + \beta_y\gamma_y + \beta_z\gamma_z = 0 \qquad \alpha_y\alpha_z + \beta_y\beta_z + \gamma_y\gamma_z = 0 \qquad \text{(h, i)}$$
$$\gamma_x\alpha_x + \gamma_y\alpha_y + \gamma_z\alpha_z = 0 \qquad \alpha_z\alpha_x + \beta_z\beta_x + \gamma_z\gamma_x = 0$$

Since six relationships (f, h) or (g, i) exist between $\alpha_x, \alpha_y, \ldots, \gamma_y, \gamma_z$, three and only three direction cosines are independent and define uniquely and completely the general rotation.

8.2   Derive the algebraic form of the stress tensors T^m and T^n, and their invariants.

Solution   If $\sigma_{xx}^n, \sigma_{yy}^n, \sigma_{zz}^n, \sigma_{xy}^n, \sigma_{yz}^n, \sigma_{zx}^n$ are the stress components in the n-system (Fig. 8-1), the stress components in the m-system (Fig. 8-2) can be shown to be

$$\sigma_{xx}^m = \alpha_x\alpha_x\sigma_{xx}^n + \alpha_x\beta_x\sigma_{xy}^n + \alpha_x\gamma_x\sigma_{xz}^n$$
$$+ \beta_x\alpha_x\sigma_{yx}^n + \beta_x\beta_x\sigma_{yy}^n + \beta_x\gamma_x\sigma_{yz}^n$$
$$+ \alpha_x\alpha_x\sigma_{zx}^n + \gamma_x\beta_x\sigma_{zy}^n + \gamma_x\gamma_x\sigma_{zz}^n \qquad \text{(a)}$$

$$\sigma_{yy}^m = \alpha_y\alpha_y\sigma_{xx}^n + \alpha_y\beta_y\sigma_{xy}^n + \alpha_y\gamma_y\sigma_{xz}^n$$
$$+ \beta_y\alpha_y\sigma_{yx}^n + \beta_y\beta_y\sigma_{yy}^n + \beta_y\gamma_y\sigma_{yz}^n$$
$$+ \gamma_y\alpha_y\sigma_{zx}^n + \gamma_y\beta_y\sigma_{zy}^n + \gamma_y\gamma_y\sigma_{zz}^n \qquad \text{(b)}$$

$$\sigma_{zz}^m = \alpha_z\alpha_z\sigma_{xx}^n + \alpha_z\beta_z\sigma_{xy}^n + \alpha_z\gamma_z\sigma_{xz}^n$$
$$+ \beta_z\alpha_z\sigma_{yx}^n + \beta_z\beta_z\sigma_{yy}^n + \beta_z\gamma_z\sigma_{yz}^n$$
$$+ \gamma_z\alpha_z\sigma_{zx}^n + \gamma_z\beta_z\sigma_{zy}^n + \gamma_z\gamma_z\sigma_{zz}^n \qquad \text{(c)}$$

$$\sigma_{xy}^m = \alpha_x\alpha_y\sigma_{xx}^n + \alpha_x\beta_y\sigma_{xy}^n + \alpha_x\gamma_y\sigma_{xz}^n$$
$$+ \beta_x\alpha_y\sigma_{yx}^n + \beta_x\beta_y\sigma_{yy}^n + \beta_x\gamma_y\sigma_{yz}^n$$
$$+ \gamma_x\alpha_y\sigma_{zx}^n + \gamma_x\beta_y\sigma_{zy}^n + \gamma_x\gamma_y\sigma_{zz}^n \qquad \text{(d)}$$

$$\sigma_{yz}^m = \alpha_y\alpha_z\sigma_{xx}^n + \alpha_y\beta_z\sigma_{xy}^n + \alpha_y\gamma_z\sigma_{xz}^n$$
$$+ \beta_y\alpha_z\sigma_{xy}^n + \beta_y\beta_z\sigma_{yy}^n + \beta_y\gamma_z\sigma_{yz}^n$$
$$+ \gamma_y\alpha_z\sigma_{zx}^n + \gamma_y\beta_z\sigma_{zy}^n + \gamma_y\gamma_z\sigma_{zz}^n \qquad \text{(e)}$$

$$\sigma_{zx}^m = \alpha_z\alpha_x\sigma_{xx}^n + \alpha_z\beta_x\sigma_{xy}^n + \alpha_z\gamma_x\sigma_{xz}^n$$
$$+ \beta_z\alpha_x\sigma_{yx}^n + \beta_z\beta_x\sigma_{yy}^n + \beta_z\gamma_x\sigma_{yz}^n$$
$$+ \gamma_z\alpha_x\sigma_{zx}^n + \gamma_z\beta_x\sigma_{zy}^n + \gamma_z\gamma_x\sigma_{zz}^n \qquad \text{(f)}$$

where by (8-9)

$$\sigma^n_{xz} = \sigma^n_{yx}, \quad \sigma^n_{yz} = \sigma^n_{zy}, \quad \sigma^n_{zx} = \sigma^n_{xz} \tag{g}$$

and analogically

$$\sigma^m_{xy} = \sigma^m_{yx}, \quad \sigma^m_{yz} = \sigma^m_{zy}, \quad \sigma^m_{zx} = \sigma^m_{xz} \tag{h}$$

Since by (g) of Problem 8.1,

$$\alpha^2_x + \beta^2_x + \gamma^2_x = 1 \tag{i}$$

two direction cosines (any two) are independent and the partial derivatives of (i) are

$$\alpha_x + \gamma_x \left(\frac{\partial \gamma_x}{\partial \alpha_x} \right) = 0 \qquad \beta_x + \gamma_x \left(\frac{\partial \gamma_x}{\partial \beta_x} \right) = 0 \tag{j, k}$$

The extreme values of σ^n_{xx} are then computed from the condition

$$\frac{\partial \sigma^m_{xx}}{\partial \alpha_x} = 0 \qquad \frac{\partial \sigma^m_{xx}}{\partial \beta_x} = 0 \tag{l, m}$$

which leads to

$$\begin{bmatrix} (\sigma^n_{xx} - \sigma^p) & \sigma^n_{xy} & \sigma^n_{xz} \\ \sigma^n_{yx} & (\sigma^n_{yy} - \sigma^p) & \sigma^n_{xz} \\ \sigma^n_{zx} & \sigma^n_{zy} & (\sigma^n_{zz} - \sigma^p) \end{bmatrix} \begin{bmatrix} \alpha_x \\ \beta_x \\ \gamma_x \end{bmatrix} = \begin{bmatrix} 0 \\ 0 \\ 0 \end{bmatrix} \tag{n}$$

and to (8-10).

Biaxial State

8.3   Given: $\sigma^p_1 = 20 \text{ kg/cm}^2$ and $\sigma^p_3 = 10 \text{ kg/cm}^2$. Using equations (8-13) to (8-16), find $\sigma^m_{xx}, \sigma^m_{zz}, \sigma^m_{xz}$ at $\omega = 30°$, and $\sigma^s_{xz,\max}$.

By (8-13)   *Solution*

$$\sigma^m_{xx} = \frac{\sigma^p_1 + \sigma^p_3}{2} - \frac{\sigma^p_1 - \sigma^p_3}{2} \cos 2\omega = \frac{20 + 10}{2} - \frac{20 - 10}{2} (0.5) = 12.5 \text{ kg/cm}^2$$

By (8-14)

$$\sigma^m_{xz} = \sigma^m_{zx} = \frac{\sigma^p_1 - \sigma^p_3}{2} \sin 2\omega = \frac{20 - 10}{2} (0.86603) = 4.3 \text{ kg/cm}^2$$

By (8-15)

$$\sigma^m_{zz} = \frac{\sigma^p_1 - \sigma^p_3}{2} + \frac{\sigma^p_1 - \sigma^p_3}{2} \cos 2\omega = \frac{20 + 10}{2} + \frac{20 - 10}{2} (0.5) = 17.5 \text{ kg/cm}^2$$

By (8-16)

$$\sigma^m_{xz,\text{max}} = \frac{\sigma^p_1 - \sigma^p_3}{2} = \frac{20 - 10}{2} = 5 \text{ kg/cm}^2$$

Note that

$$J_1 = \sigma^p_1 + \sigma^p_2 = \sigma^m_{xx} + \sigma^m_{zz} = 30 \text{ kg/cm}^2$$

is the invariant of the stress tensor of the biaxial state.

8.4   Using Mohr's circle, Table 8-2, check the results of Problem 8.3.

Solution   Referring to Table 8-2:

1.   Plot σ^p_1 and σ^p_3 on the σ_N axes and designate the endpoints as 1 and 2.
2.   Draw a circle through 1 and 2 with center c on the σ_N axis (Mohr's circle).
3.   Find the points 3 and 4 given by $2\omega = 60°$ and $\pi + 2\omega = 240°$ respectively.
4.   Scale $\sigma^m_{xx}, \sigma^m_{zz}, \sigma^m_{zx}$ from the circle.
5.   Find the points 5 and 6 given by $\pi/2$ and $3\pi/2$ respectively.
6.   Scale $\sigma^s_{xx} = \sigma^s_{zz}$ and $\sigma^s_{xz,\text{max}}$ from the circle.

Strength Theories and Laboratory Tests

8.5   A series of triaxial tests were performed on samples of a silty-clay soil. After plotting the results, the values of cohesion c and angle of internal friction ϕ were found to be 1100 lb/in.^{2} and 18 degrees respectively.   Determine the value of the vertical pressure that would cause failure of this soil if the confining pressure is 500 lb/ft^2.

Solution   Assuming that equation (8-19) is valid in this case,

$$\sigma^p_1 = 2c \tan \left(\frac{\pi}{4} + \frac{\phi}{2} \right) + \sigma^p_3 \tan^2 \left(\frac{\pi}{4} + \frac{\phi}{2} \right)$$

Substituting in the preceding equation for the given values of c, ϕ, and $\sigma^p_3 = 500$ lb/ft^2,

$$\sigma^p_1 = 2(1,100) \tan \left(45° + \frac{18°}{2} \right) + 500 \tan^2 \left(45° + \frac{18°}{2} \right) = 3027 + 947$$

$$= 3924 \text{ lb/ft}^2$$

8.6   Describe the direct shear test and outline the laboratory procedure for conducting this test.

Answer   The direct shear test apparatus is shown diagrammatically in Fig. P-8.6a. It consists of a metal box, split horizontally at the center of the soil specimen. The usual size of the sample is 6 by 6 cm. However, for testing granular soils or gravel material, it is necessary to use a larger box, generally 12 by 12 in.

A vertical pressure is applied to the sample by means of weights. The sample is sheared by horizontally moving the upper and lower halves of the box relative to each other at a constant rate of strain, usually about 0.05 in./min. for the undrained

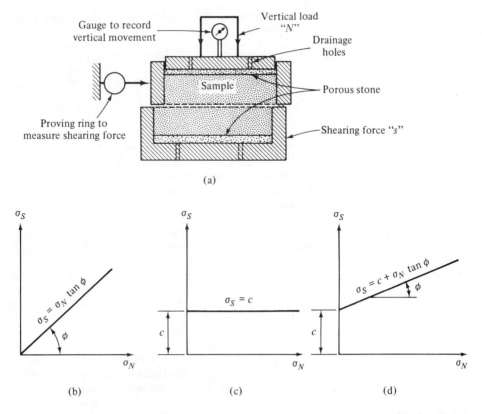

Figure P-8.6

test. When testing clay samples in this apparatus, an undisturbed sample is prepared by a special cutter. But when testing cohesionless soils (e. g., sands), the material is compacted in the box to the desired density.

The shear box is generally kept inside a container so that it can be filled with water to allow saturation of the sample. The two porous plates at both ends of the sample, however, permit drainage of the sample in a drained test. Volume changes of the soil sample during the the testing can be read on a dial gage measuring the vertical movements of the upper loading plate. Three types of tests can be performed by this apparatus: (a) undrained test, in which the shearing starts immediately after applying the vertical load so as not to give any time for consolidation (test completed in about 5 to 10 minutes); (b) consolidated-undrained test, in which the specimen is allowed to consolidate fully under the applied vertical load. Afterward the sample is sheared quickly (in about 5 to 10 minutes); (c) drained test, in which the sample is sheared very slowly. The specimen is first allowed to consolidate under vertical pressure and then sheared at a very slow rate (from 2 to 5 days) to allow complete dissipation of the generated pore-water pressure.

In this test, a number of identical specimens are tested under increasing normal loads and the required maximum shear force is recorded and then plotted on a graph versus the normal stress. Examples of the results of the direct shear test on three types of soils (cohesionless, purely cohesive, and c-ϕ soils respectively) are shown in parts (b), (c), and (d) of Fig. P-8.6.

8.7   List the advantages of the direct shear test.

Answer   Advantages and disadvantages of the direct shear test are listed in Table P-8.7.

TABLE P-8.7

Advantages	Disadvantages
1. The test is simple and can be performed rather rapidly compared to a triaxial compression test (see Problem 8.10).	1. There is usually a nonuniformity of shear-stress distribution over the surface of the sample (stress is maximum at the edges and minimum at center).
2. Quick drainage of pore water in sample is usually easy to achieve because of the thin depth of the sample.	2. The area along the potential surface of sliding gradually decreases with the progress of the test.
3. The test is most suitable for cohesionless soils.	3. Sometimes it is difficult to control the drainage of the specimen (e. g., it would be very difficult to prevent the escape of pore water from a highly permeable soil if an undrained test for such a soil is desired).
	4. The shear failure plane is predetermined and may not necessarily be the weakest one.
	5. The lateral restraint by the side walls of the shear box will have a definite effect on the results of the test. Such lateral restraint may not exist in the actual foundation in the field.

8.8   A number of samples of silty-clay soil were tested in a shear box, 36 cm² in area, under undrained conditions. The results of the tests are shown in Table P-8.8. Determine the values of apparent angle of shearing resistance ϕ and the apparent cohesion c.

TABLE P-8.8

Normal load, lb	5	15	24	35	45
Maximum shear force, lb	10.5	14.0	17.5	22.0	26.0

Solution   The results are plotted as shown in Fig. P-8.8. By direct measurement from the graph, the apparent cohesion C is 8 lb and angle of internal friction ϕ is 20 degrees. Thus the unit cohesion is

$$c = \frac{C}{\text{area of sample}} = \frac{8}{36} = 0.22 \text{ lb/cm}^2$$

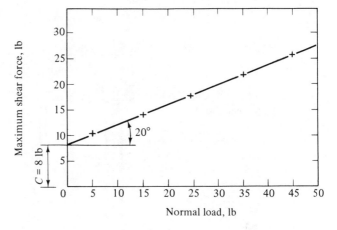

Figure P-8.8

8.9 Describe the ring shear test.

Figure P-8.9 illustrates the main features of the direct double-shear ring apparatus. *Answer*
The shearing force P_s is applied to the central movable ring so as to produce shearing
along a cylindrical surface A. The test usually yields a low value of shearing resistance because not all of the normal force P_n would be transmitted to the planes A
and part of it is taken by friction along the walls of the outer ring. On the other
hand, expansion of dense sands might be partially prevented for the same reason,
which would cause too high a value of shearing resistance.

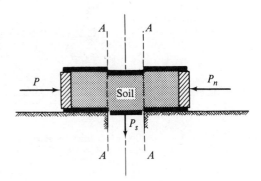

Figure P-8.9

The apparatus is generally designed so that the inner tube of a special sample is
formed by rings that will fit directly into the shearing device, without having to
transfer the soil from them, thus reducing the disturbance of the sample.

8.10 Describe briefly the triaxial shear test apparatus and procedure.

Vertical load

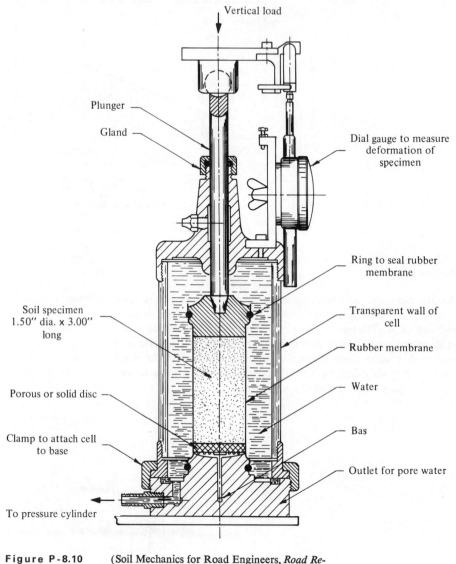

Plunger

Gland

Dial gauge to measure
deformation of
specimen

Ring to seal rubber
membrane

Soil specimen
1.50″ dia. x 3.00″
long

Transparent wall of
cell

Rubber membrane

Porous or solid disc

Water

Clamp to attach cell
to base

Bas

Outlet for pore water

To pressure cylinder

Figure P-8.10   (Soil Mechanics for Road Engineers, *Road Research Laboratory, H.M.S.O.*, England, 1956,
p. 363.)

Answer   The apparatus is shown diagrammatically in Fig. P-8.10. In this test, a cylindrical
specimen of soil, usually 3 in. long by 1.5 in. wide, is subjected to three compressive
stresses at right angles to one another, and one of these stresses is increased until the
specimen fails in shear. The basic difference between this test and that of the direct
shear box (Problem 8.6) is that in this test the shear failure plane is not predeter-
mined.

In the usual types of triaxial tests, the specimens are subjected to constant radial stresses generated by fluid pressure and an increasing axial stress generated by a loading system. A number of identical specimens must be tested, with each being subjected to a different constant radial (confining) pressure. The axial stress is then increased until each specimen fails. It should be noted that the load shown on the proving ring gage dial indicates the deviator stress or principal stress difference, $(\sigma_1^p - \sigma_3^p)$. For each test, the total axial pressure at failure σ_1^p and the lateral pressure σ_3^p are plotted as shown in Fig. 8.4 and the Mohr circle is drawn. After the Mohr's circles for several tests are drawn, the failure envelope is drawn as a tangent to these circles. From such a plotting, the apparent cohesion and angle of shearing resistance are determined (see Fig. 8.4).

For very coarse grain soils, large triaxial cells, up to 8 in. long and 4 in. wide, are used. Since the radial pressure is applied to the specimen by fluid pressure (water or glycerine), the specimen must be enveloped in an impermeable rubber membrane and contained in a pressure cell (usually composed of a transparent material like plexiglass so that the specimen can be kept under observation throughout the test). When pore-pressure measurements are desired, porous ends are used and the outlet is connected to the pore-pressure measuring apparatus. The specific test procedure selected depends on the purpose for which the triaxial strength is used. The following procedures are available: (a) quick or undrained test, (b) consolidated-quick or consolidated-undrained test, and (c) slow or drained test (see Problems 8.14, 8.15, 8.16).

Strength of Cohensionless Soils

8.11 A saturated specimen of cohesionless soil was tested in a drained triaxial compression test. The sample failed in shear under a deviator stress of 76 lb/in.^{2} when the confining pressure was 21.5 lb/in^2. Find the effective angle of shearing resistance of sand and the approximate orientation of the failure plane with reference to the horizontal. Also, determine the deviator stress and the major principal stress at failure for another identical sample of the same soil when tested in the same triaxial test under a confining pressure of 30 lb/in^2.

In drained triaxial tests, the effective stresses are equal to the total stresses. Designating the first sample as A,   *Solution*

$$\sigma_{3A}^p = 21.5 \text{ lb/in.}^2$$

Deviator stress

$$\sigma_{dA} = \sigma_{1A}^p - \sigma_{3A}^p = 76 \text{ lb/in.}^2$$

Hence

$$\sigma_{1A}^p = \sigma_{dA} + \sigma_{3A}^p = 21.5 + 76.0 = 97.5 \text{ lb/in.}^2$$

A Mohr's circle is then drawn as shown in Fig. P-8.11. Since the soil is cohesionless, the failure envelope passes through the origin of stress. It is also drawn tangent to the Mohr circle. From this figure, the inclination of the failure envelope, $\phi_1 = 40°$.

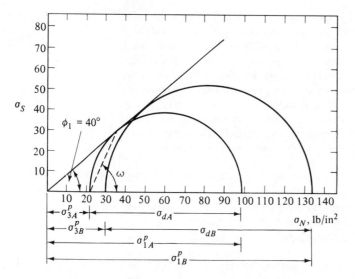

Figure P-8.11

The angle of inclination of the failure surface to the horizontal is ω (Fig. P-8.11),

$$\omega = \left(45° + \frac{\phi}{2}\right) = \left(45 + \frac{40}{2}\right) = 75°$$

The Mohr's circle of the second specimen B is then drawn with its center on the σ_N axis, intersecting it at a point $\sigma_{3B}^p = 30$ lb/in.^{2} and touching the failure envelope. From Fig. P-8.11,

$$\sigma_{dB} = 103 \text{ lb/in.}^2 \qquad \sigma_{1B}^p = 133 \text{ lb/in.}^2$$

8.12 Consolidated-undrained triaxial tests were conducted on samples of saturated silty-sand (cohesionless) soil. Pore-pressure measurements and other test results were recorded during the test, as shown in Table P-8.12a. Determine the value of the angle of internal friction for this soil in terms of total and effective stresses.

Solution For these two tests, the major and minor principal stresses are calculated as shown in Table P-8.12b. Using the calculated values of $\sigma_1^p, \sigma_3^p, \sigma_1^{\bar{p}}, \sigma_3^{\bar{p}}$ shown in the table, the Mohr circles and failure envelopes in terms of total and effective stresses are drawn as shown in Fig. P-8.12. From this figure, the angle of shearing resistance in terms of total stresses for the consolidated-undrained test is designated as ϕ_u, and that in terms of effective stresses as ϕ_1. When measured (Fig. P-8.12), we have

$$\phi_u = 15° \quad \text{and} \quad \phi_1 = 22°$$

8.13 If a third soil sample in Problem 8.12 were subjected to a consolidation pressure of $P_c = 60$ lb/in.^{2}, which was kept constant afterward during the shear test, what would be the deviator stress at failure for this sample?

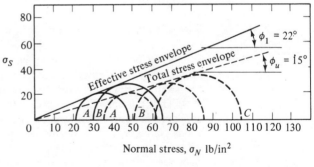

Figure P-8.12

To solve this problem, a Mohr's circle C is drawn in Fig. P-8.12 with its center   *Solution*
on the σ_N axis passing through a point $\sigma_3^p = 60 \ \text{lb/in.}^2$ and tangential to the total
stress failure envelope.

By direct measurement from Fig. P-8.12,

$\sigma_1^p \cong 104 \ \text{lb/in.}^2$

$\sigma_d \cong \sigma_1^p - \sigma_3^p = 104 - 60 = 44 \ \text{lb/in.}^2$

TABLE P-8.12a

Test Number	Consolidation Pressure P_c	Confining Pressure at Failure σ_3^p	Deviator Stress at Failure σ_d	Pore Pressure at Failure u_w
1	35	35	26	14.40
2	50	50	35	20.70

All pressures and stresses in pounds per square inch.

TABLE P-8.12b

Test Number	σ_3^{p*}	σ_1^{p*}	σ_{d*}	u_{w*}	$\sigma_3^{\bar{p}*}$	$\sigma_1^{\bar{p}*}$	σ_{d*}
1	35	61	26	14.40	20.60	46.60	26
2	50	85	35	20.70	29.30	64.30	35

*Given in pounds per square inch.

Strength of Cohesive Soils

8.14 Describe the consolidated-drained shear test and its principal applications.

Answer In this test (which is also known as the "slow test"), drainage is permitted to take place throughout the test, during the applications of both major and minor principal stresses in the triaxial test (or the normal and shearing forces in the direct shear box test). Thus full consolidation occurs for the sample and no pore-water pressure is developed at any stage of the test. A triaxial test under these conditions may last for a long period of time (sometimes for more than one week). Difficulties may arise in maintaining a constant cell pressure during such a long testing period, and a special elaborate control may have to be utilized. A self-compensating mercury control system was originally developed for this purpose by the Imperial College in London (1956).

The results of this consolidated-drained test are used to determine strength of soils in connection with field problems where the field stresses develop within the soil mass sufficiently slowly for all changes in moisture content to take place. An example of such a case is the determination of the final bearing capacity of a soil mass under the foundation of a structure that is erected more slowly than the soil consolidates. It is also used for determining strength of sandy soils or for clay embankments in which drainage channels are embedded.

8.15 Describe the consolidated-undrained shear test and its principal applications.

Answer In this test (which is also known as the consolidated-quick test), drainage of the sample tested is only permitted under the initially applied confining pressure in the triaxial test or under the initially applied normal load in the direct shear test, and full primary consolidation is allowed to take place. Afterward no drainage is allowed during the subsequent applications of the normal or shear stresses.

The results of this test are generally used in estimating soil strength in field situations where the soil has consolidated under the pressure of the foundation during construction or under its own weight, and which is then followed by quick increase in loads, causing a rapid change in critical stresses during which no further change in water content can take place. Examples of such cases are the stability calculations against failure by shear in consolidated dams, slopes, and other earth structures of cohesive soils under conditions of rapid drawdown of water, where water has no time to drain out of the voids. Other examples would include cases of consolidated clay soils serving as foundations for grain elevators under conditions of quick unloading.

8.16 Describe the unconsolidated-undrained shear test and its principal applications.

Answer This test is also known as the quick or immediate test, for one test may take no more than a few minutes. No drainage is permitted during the entire test.

The results of such a test can be used in estimating the strength of soils for field problems where critical stresses develop in a saturated soil mass too rapidly for any water content to change appreciably. An example of such a case is the analysis of slope stability problems ($\phi = 0$ analysis; see Chapter 9), for saturated clays and for foundations on clay soils where drainage is too slow.

8.17 Describe the behavior of normally consolidated samples of clay soils when tested in a consolidated-undrained shear test.

Specimens in a consolidated-undrained test are first consolidated under a certain *Answer* cell pressure in triaxial apparatus (or under a normal pressure in a shear box) and then sheared by increasing the axial stress under the undrained condition, but maintaining the same cell pressure (or normal pressure in case of shear box) under which the sample was previously consolidated. When test results for this test are plotted in terms of total and effective stresses, they will show trends similar to those illustrated in Fig. P-8.17.

After consolidation under cell pressure is complete, no excess pore-water pressure will be present inside the sample. However, when the specimen is subjected to the subsequent increase of axial stress, under undrained conditions, an increase in pore-water pressure u_w will occur. Thus the effective stress $\sigma_1^{\bar{p}}$ and $\sigma_3^{\bar{p}}$ at failure will be less than the corresponding total stresses σ_1^{p} and σ_3^{p} by the value of u_w. But the difference $\sigma_1^{p} - \sigma_3^{p}$ will be equal to $\sigma_1^{\bar{p}} - \sigma_3^{\bar{p}}$, and hence the diameters of the Mohr's circles in terms of total stress are the same as those for effective stress, except that the effective stress circles are shifted to the left, as shown in Fig. P-8.17.

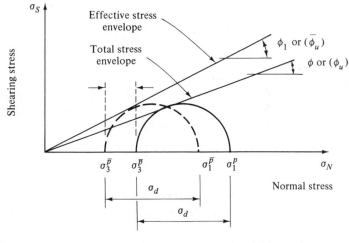

Figure P-8.17

It can also be clearly seen from Fig. P-8.17 that the effective angle of shearing resistance ϕ_1 is greater than the apparent angle ϕ (or ϕ_u). Furthermore, it can be seen that the cohesion intercept for normally consolidated soil is zero both in terms of total and effective stresses [see equation (8.12)].

8.18 Describe the behavior of preconsolidated samples of cohesive soils when tested in a consolidated-undrained shear test.

The behavior of preconsolidated clay samples in a consolidated-undrained triaxial *Answer* shear test is illustrated in Fig. P-8.18. The total stress failure envelope is usually

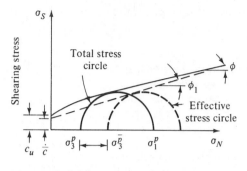

Figure P-8.18

curved in the initial portion and up to the preconsolidation pressure, after which it plots as a straight line. As shown, the failure envelope also has a cohesion intercept. The pore-water pressure developed with increasing axial load in a consolidated-undrained test will be negative, up to a cell pressure equal to that of the preconsolidation pressure; and thus the effective stress Mohr circle will be shifted toward the right relative to the total stress circle, as shown in the figure.

It can be also observed that the effective cohesion $\bar{c}$ is smaller than the apparent cohesion c_u. The shear failure envelope for the effective stress up to the preconsolidation pressure may be expressed as

$$S = \bar{c} + \bar{p} \tan \phi_1 \qquad \text{(a)}$$

which is the same as (8-23), and in which $\bar{p}$ is the effective pressure under which the soil was consolidated prior to shearing. Above the preconsolidation pressure, where soil would be considered as a normally consolidated material, equation (a) reduces to

$$S = \bar{p} \tan \phi_1 \qquad \text{(b)}$$

8.19 Describe the behavior of normally consolidated and preconsolidated samples of cohesive soils when tested in a drained shear test

Answer Here the slow process of loading does not allow excess pore-water pressure u_w to build up during the test. Hence the total stresses are the same as the effective stresses. Figure P-8.19 shows the failure envelope for the Mohr circles plotted from the results of the consolidated-drained tests, where

$\bar{c}_d$ = effective cohesion intercept for drained test

ϕ_{1d} = effective angle of shearing resistance for drained test

The shear strength equation for consolidated drained test is expressed as

$$S = \bar{c}_d + \bar{p} \tan \phi_{1d}$$

in which $\bar{c}_d = 0$ for the normally consolidated condition.

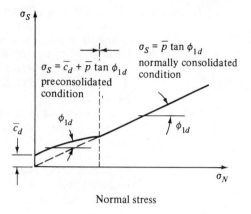

8.20 Describe the behavior of samples of cohesive soils tested in a consolidated-un-drained triaxial test, if the samples are first consolidated under the same cell pressure and then sheared under undrained conditions with different cell pressures.

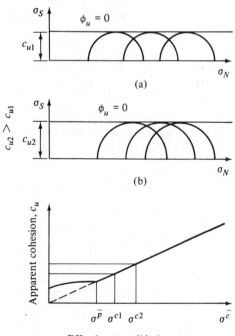

Answer This condition is different from the one described in Problem 8.17 in that the cell pressure used in consolidating the samples was not maintained at a constant level during the subsequent shearing of the samples under the increasing axial pressure. In such a case, the deviator stress at failure is independent of the cell pressure, and the test results can be represented by a series of Mohr circles with equal diameters. The apparent angle of internal friction in this case is zero ($\phi_u = 0$), as shown in Fig P-8.20a.

If another set of samples of the same soil is first consolidated under a higher cell pressure than that used in the first case, and is then sheared with variable cell pressures under undrained conditions, ϕ_u will still be equal to zero. But the apparent cohesion c_{u2} (Fig. P-8.20b) will be greater than that in the first case, or $c_{u2} > c_{u1}$. This situation can be explained by the fact that, under higher consolidation pressure, void ratios will be smaller, thus resulting in a higher undrained strength. The relationship between apparent cohesion c_u and effective consolidation pressure is shown in Fig. P-8.20c.

8.21 Explain the practical significance of the relationship discussed in Problem 8.20.

Answer A relationship similar to that outlined in Problem 8.20 was found to exist in the consolidated-undrained shear strength of natural clays. For uniform, normally consolidated clays, the undrained shear strength $s = c_u$ was found to increase approximately linearly with the effective vertical overburden pressure $\bar{p}$. According to Skempton (1957), the ratio $(c_u/\bar{p})$ was found to be closely related to the plasticity index (PI) in a normally consolidated clay. The approximate correlation may be expressed as

$$\frac{c_u}{\bar{p}} = 0.11 + 0.0037 \, (\text{PI})$$

8.22 Describe the behavior of samples of cohesive soil when tested in unconsolidated, undrained (quick) shear test.

Answer Typical results of an undrained (quick) shear test on saturated cohesive soils are shown in Fig. P-8.22. It can be seen from this figure that since no drainage is allowed,

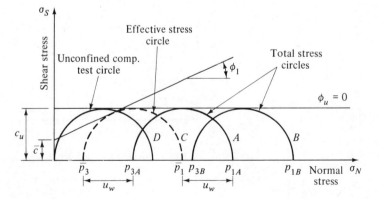

Figure P-8.22

the increase in cell pressure σ_3^p only results in an equal increase in pore-water pressure u_w, and the subsequent increase in the total major principal stress σ_1^p also results in the same change in pore-water pressure. However, the major and minor effective principal stresses and the deviator stresses remain unchanged and are independent of the cell pressure. Hence

$$(\sigma_1^p)_A - (\sigma_3^p)_A = (\sigma_1^p)_B - (\sigma_3^p)_B = p_1 - p_3 = \bar{p}_1 - \bar{p}_3 \qquad (a)$$

Thus all the total stress circles and the only one corresponding effective stress circle will be of the same diameter. No effective failure envelope can possibly be established from this test, for there is only one effective stress circle. The total stress envelope in terms of total stress is horizontal; that is,

$$\phi_u = 0 \quad \text{and} \quad S = c_u = \frac{\sigma_1^p - \sigma_3^p}{2} \qquad (b)$$

It should, however, be noted that if fissured clays are tested in this quick test, the deviator stress $(\sigma_1^p - \sigma_3^p)$ will not be independent of the cell pressure if it is less than the overburden pressure that existed in the field over the sample before it was removed from the ground. Also, the angle of internal friction ϕ_u will be greater than zero. When the true strengths of fissured clays are determined in this type of test, it is therefore necessary that they be subjected to a cell pressure exceeding the overburden pressure in the field.

8.23 Explain the value of the unconfined compression test as a laboratory shear strength test for soils.

The unconfined compression test may be considered as a special form of an uncon- *Answer*
solidated-undrained (quick) triaxial shear test in which the cell pressure $\sigma_3^p = 0$. Circle D on Fig. P-8.22 represents the Mohr circle for the unconfined compression test. The test is based on the assumption that there is no moisture loss during the test.

The unconfined compression test is used in evaluating the strength of remolded and undisturbed cohesive soils. Cohesionless soils (sands and gravels) cannot be tested in this type of test because they cannot be formed in unsupported cylinders or prisms. Since the angle of internal friction ϕ_u of plastic saturated clays can be assumed to be approximately equal to zero, equation (b) of Problem 8.22 reduces to

$$S = c \quad \text{and} \quad \sigma_1^p = 2c \quad \text{or} \quad c = \frac{\sigma_1^p}{2} = S$$

The inclination of the shear failure plane to the horizontal is

$$\omega = \left(\frac{\pi}{2} + \frac{\phi_u}{2} \right)$$

but since $\phi_u = 0$, then $\omega = 45°$.

8.24 Define the pore-pressure coefficients as proposed by Skempton (1954) and explain their significance in a soil system.

Answer   If a sample of clay soil is tested in an undrained triaxial test and subjected to a cell pressure of $\Delta\sigma_3^p$, then the pore-water pressure induced in the sample will also change according to the relationship

$$\Delta u_{w(1)} = B(\Delta\sigma_3^p)$$

Hence

$$B = \frac{\Delta u_{w(1)}}{\Delta\sigma_3^p} \tag{a}$$

Also, when the test continues and the sample, still under an undrained condition, is subjected to a deviator stress of $(\Delta\sigma_1^p - \Delta\sigma_3^p)$, the pore-water pressure change induced in the sample will be

$$\Delta u_{w(2)} = \bar{A}(\Delta\sigma_1^p - \Delta\sigma_3^p) \tag{b}$$

However, the total pore-water pressure change is

$$\Delta u_w = \Delta u_{w(1)} + \Delta u_{w(2)}$$

or

$$\Delta u_w = B\Delta\sigma_3^p + \bar{A}(\Delta\sigma_1^p - \Delta\sigma_3^p) = B[\Delta\sigma_3^p + A(\Delta\sigma_1^p - \Delta\sigma_3^p)] \tag{c}$$

where

$$\bar{A} = AB$$

Both $\bar{A}$ and B are known as Skempton's pore-water pressure parameters. They are empirical parameters and can only be obtained experimentally from the results of undrained triaxial tests in which pore-water pressures are measured.

The parameter B is most influenced by the degree of saturation of the soil as follows:

$B = 1.0$ for a fully saturated soil,

$B = $ zero   for a dry soil,

$0 < B < 1$ for partially saturated soils

The parameter A, however, is influenced by the stress condition, the rate of consolidation (or strain), and the stress history that the sample has been subjected to, where

$A = 1.5$ for a highly sensitive clay

$A = -0.5$ for a heavily preconsolidated clay

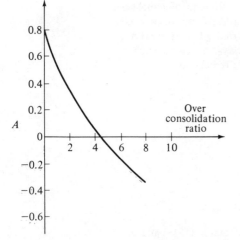

Figure P-8.24

Figure P-8.24 shows the relationship between the parameter A and the overconsolidation ratio. The overconsolidation (or preconsolidation) ratio is defined as the ratio of preconsolidation pressure and the cell pressure used in the test. When the overconsolidation ratio is 1.0, the soil is considered normally consolidated. There is a growing tendency at present to use the effective stress principles and pore-water pressure parameters in the design of many earth structures.

8.25   A shear test was conducted in a layer of clay soil by pressing a vane of 4 in. height and 2.5 in. diameter into the clay layer. A torque was then gradually applied, and failure took place at a torque pressure of 400 lb in. What is the approximate value of the shear strength of this soil at a horizontal plane?

By (8.25)   *Solution*

Torque $T = c\pi \left(\dfrac{d^2 h}{2} + \dfrac{d^3}{6} \right)$

or

$400 = c\pi \left[\dfrac{(2.5)^2(4)}{2} + \dfrac{(2.5)^3}{6} \right]$

Hence the cohesion

$c = 8.43$ lb/in.^{2}

which is considered the same as the strength S, or

$S = c = 8.43$ lb/in.^{2}

Relationships Between Strength Tests

8.26   A triaxial shear strength test was conducted on a sample of cohesive soil. The test yielded the following results: major principal stress $\sigma_1^p = 900$ lb/in.^{2}, minor principal stress $\sigma_3^p = 200$ lb/in.^{2}, and the angle of inclination of the failure surface to the horizontal $= 60°$. Present all the information pertaining to this problem in the form of Mohr's stress diagram. From this stress diagram, determine the normal stress σ_{xx}^m, the shear stress σ_{zz}^m, and the resultant stress σ_r^m on the rupture plane through a point and its angle of inclination with the rupture surface.

The Mohr's stress diagram for this condition is shown in Fig. P-8.26. From this dia-   *Solution*
gram, and by direct scaling off the figure, the required parameters can be determined.

$\sigma_1^p = 900$ lb/in.^{2}, $\sigma_3^p = 200$ lb/in.^{2}, $\sigma_{xx}^m = 375$ lb/in.^{2}

$\sigma_r^m = 480$ lb/in.^{2}, $c = 95$ lb/in.^{2}, $\phi = 30°$

$p_2 = 175$ lb/in.^{2}, $\beta = 39°$ angle with rupture surface $= 90 - \beta = 51°$

All quantities are indentified in Fig. P-8.26, as well as in Fig. 8.3 and Table 8-2.

8.27   Determine the same quantities as in Problem 8.26 analytically, using the relationships in Sections 8.2 and 8.3. Compare the values obtained by analytical ap-

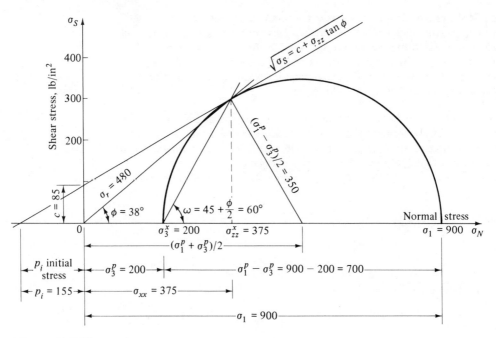

Figure P-8.26

proach to those obtained by direct measurements from the Mohr's diagram in Problem 8.26.

Solution   By (8-13), the normal stress is

$$\sigma_{xx}^m = \frac{\sigma_1^p + \sigma_3^p}{2} - \frac{\sigma_1^p - \sigma_3^p}{2}\cos 2\omega = \frac{900 + 200}{2} - \frac{900 - 200}{2}\cos 120°$$
$$= 550 - 175 = 375 \text{ lb/in.}^2$$

By (8-14), the shear stress is

$$\sigma_{zx}^m = \frac{\sigma_1^p - \sigma_3^p}{2}\sin 2\omega = \left(\frac{900 - 200}{2}\right)(\sin 120°) = 303 \text{ lb/in.}^2$$

The resultant stress is

$$\sigma_r^m = \sqrt{(\sigma_{xx}^m)^2 + (\sigma_{zx}^m)^2} = \sqrt{(375)^2 + (303)^2} = 483 \text{ lb/in.}^2$$

The angle of inclination of the resultant stress to the horizontal is

$$\beta = \arctan^{-1}\frac{\sigma_{zx}}{\sigma_{xx}} = \arctan^{-1}\frac{303}{375} = 38°57'$$

Hence the angle of inclination of σ_r^m with the shear plane is

$(90 - \beta) = 90° - 38°57' = 51°3'$

The relationship of ω and ϕ shown in Fig. P-8.26 is

$$\omega = 45° + \frac{\phi}{2} = 60°$$

from which

$$\phi = 2(60° - 45) = 30°$$

By (8-23), the shear strength is

$$S = \sigma_{zx}^m = c + \sigma_{xx}^m \tan \phi$$

or

$$303 = c + 375(\tan 30°)$$

Thus

$$c = 303 - 375(0.577) = 87 \text{ lb/in.}^2$$

Once c and ϕ are known, the initial stress becomes

$$p_2 = \frac{c}{\tan \phi} = \frac{87}{\tan 30°} = \frac{87}{0.577} = 151 \text{ lb/in.}^2$$

All values of parameters obtained by analytical procedure in this problem are very close to those obtained by direct measurements from the Mohr's diagram in Problem 8.26.

8.28   A consolidated-undrained triaxial test, with pore-water pressure measured, was conducted on a sample of soil from a clay fill. From the results of the test, the effective cohesion $\bar{c}$ was found to be 4.0 lb/in.^{2} and the effective angle of internal friction ϕ_1 to be 30 degrees. If the natural density of the fill is 118 lb/ft^3, what would be the value of the shear strength of the embarkment material at a depth of 60 ft below the surface of the embankment if the pore-water pressure at this depth is 25 lb/in.^{2}.

The vertical pressure resulting from the weight of embankment at a depth of 60 ft   *Solution*
is

$$p = d\gamma_m = (60)(118) = 7080 \text{ lb/ft}^2 \quad \text{or} \quad 49.1 \text{ lb/in.}^2$$

The effective pressure at this depth is

$$\bar{p} = p - u_w = 49.1 - 25.0 = 24.1 \text{ lb/in.}^2$$

By (8.23), the shear strength is

$$S = \bar{c} + \bar{p} \tan \phi_1 = 4.0 + (24.1)(\tan 30°) = 4.0 + 13.9 = 17.9 \quad \text{or} \quad 18 \text{ lb/in.}^2$$

8.29   Table P-8.29a shows the results from undrained shear box tests conducted on specimens of a sandy-clay soil; each has a cross-sectional area of 36 cm^2(or 5.58 in.^{2}). If a sample of the same soil is tested in a triaxial compression test with a confining pressure σ_3^p of 22.5 lb/in.^{2}, what is the total normal stress σ_1^p under which the sample would fail ?

TABLE P-8.29a

Vertical load, lb	231.6	209.2	117.2	36.3
Shearing force at failure, lb	139.5	131.1	97.7	69.8

Solution   From Table P-8.29a the relationship between the normal stress σ_N and the shear σ_S can be established by dividing values of vertical load and shearing force by the area of the sheared sample—that is, 5.58 in^2. The results are tabulated in Table P-8.29. By plotting these results, as shown in Fig. P-8.29, we find that

$$c_u = 10 \text{ lb/in}^2 \quad \text{and} \quad \phi_u = 20°$$

The second step is to plot the Mohr circle for the triaxial test at σ_3^p of 22.5 lb/in.^{2} To locate the center of this circle, draw a line from σ_3^p at an inclination of $(\pi/4 + \phi_u/2)$ or at 55 degrees to cut the failure envelope at A.

TABLE P-8.29b

Normal stress, lb/in.^{2}	41.5	37.5	21.0	6.5
Shear stress, lb/in.^{2}	25.0	23.0	17.5	12.5

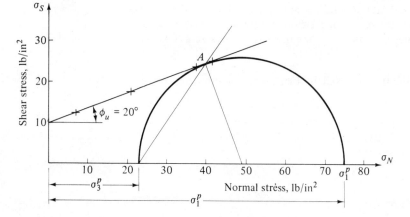

Figure P-8.29

Then draw a line AO perpendicular to the failure envelope that determines the center of the circle at point O (Fig. P-8.29). The total normal stress σ_1^p will then be fixed at the point where the circle intercepts the normal stress axis; and it is found to be 75 lb/in.²

8.30   A layer of silty-clay soil is located at a depth of 60 ft below ground surface. The overlying soil has a bulk density of 124 lb/ft³, and the water table is at the surface of the ground. Undisturbed samples of the silty-clay layer were tested in a triaxial shear test under drained and undrained conditions, and the following data were obtained: $c_d = 7$ lb/in.², $\phi_1 = \phi_d = 25°$, and $c_u = 9$ lb/in.², $\phi_u = 15°$. Determine the shearing resistance of the silty-clay layer on a horizontal plane under the following conditions:

(a)   When ground is subjected to quick loading where shear stresses within the soil mass would be expected to build up rapidly.
(b)   When ground is subjected to gradual loading where shear stresses within the soil mass would be expected to build up slowly and gradually.

Submerged density of overburden soil is   *Solution*

$$\gamma_{sub} = \gamma_m - \gamma_0 = 124 - 62.4 = 61.6 \text{ lb/ft}^3$$

Total normal pressure at depth of 60 ft is

$$p = z\gamma_m = (60)(124) = 7440 \text{ lb/ft}^2 \quad \text{or} \quad 51.6 \text{ lb/in.}^2$$

Effective pressure at depth of 60 ft is

$$\bar{p} = z\gamma_{sub} = (60)(61.60) = 3696 \text{ lb/ft}^2 \quad \text{or} \quad 25.6 \text{ lb/in.}^2$$

(a)   For conditions of a rapid buildup of shear stresses within the soil mass, the total pressure and soil properties under undrained conditions are used. Hence, by (8-23),

shear strength $S = c_u + p \tan \phi_u = 9 + (51.6)(\tan 15°)$
$$= 9 + 13.8 = 22.8 \text{ lb/in.}^2$$

(b)   For conditions of gradual and slow buildup of shear stresses within the soil mass, the effective pressure and soil properties under drained conditions are used. Hence, by (8-23),

shear strength $S = c_d + \bar{p} \tan \phi_d = 7 + (25.6)(\tan 25°)$
$$= 7 + 11.9 = 18.9 \text{ lb/in.}^2$$

SUPPLEMENTARY PROBLEMS

Triaxial State

8.31   Evaluate the coefficients of Π^{mn} in (8-5). Given: $\theta_x = 45°, \theta_y = 0, \theta_z = 90°$.

Answer Use the triple matrix product of Problem 8.1.

$$\Pi^{mn} = \begin{bmatrix} 0 & 1 & 0 \\ -1 & 0 & 0 \\ 0 & 0 & 1 \end{bmatrix}\begin{bmatrix} 1 & 0 & 0 \\ 0 & 1 & 0 \\ 0 & 0 & 1 \end{bmatrix}\begin{bmatrix} 1 & 0 & 0 \\ 0 & 0.707 & 0.707 \\ 0 & -0.707 & 0.707 \end{bmatrix}$$

8.32   Given: $\sigma_1^p = 30 \text{ kg/cm}^2$, $\sigma_2^p = 20 \text{ kg/cm}^2$, $\sigma_3^p = 10\text{kg/cm}^2$. Find the extreme shearing stresses in the given soil system.

Answer $\sigma_{1,23}^s = \pm 5 \text{ kg/cm}^2$, $\sigma_{2,31}^s = \pm 10 \text{ kg/cm}^2$, $\sigma_{3,12}^s = \pm 5 \text{ kg/cm}^2$

Biaxial State

8.33   Derive the equation of Mohr's circle.

Answer Equation (8-18)

8.34   Given: $\sigma_1^p = 30 \text{ kg/cm}^2$, $\sigma_3^p = 10 \text{ kg/cm}^2$. Find the plane and magnitude of extreme shearing stress.

Answer $\omega = \dfrac{\pi}{4}, \dfrac{5\pi}{4}$, $\sigma_{xz,\text{max}}^s = 10 \text{ kg/cm}^2$

Strength Theories and Laboratory Tests

8.35   The following results were obtained from undrained shear box tests on samples of a silty-clay soil. Determine the values of apparent cohesion and angle of internal friction.

TABLE P-8.35

Normal pressure, lb/in.^{2}	20	45	60	95
Shear strength, lb/in.^{2}	15	22	26	35

Answer $c_u = 10 \text{ lb/in.}^2$, $\phi_u = 15°$

8.36   If another specimen of the soil in Problem 8.35 was tested in an undrained triaxial shear test with a cell pressure of 50 lb/in.^{2}, using a Mohr circle, determine the value of the total axial stress at which failure of the sample would be expected.

Answer Total normal stress $\sigma_1^p = 110 \text{ lb/in.}^2$

8.37   Solve Problem 8.36, using the analytical relationship expressed in (8-19).

Answer Total normal stress $\sigma_1^p = 110.95 \text{ lb/in.}^2$

8.38   A dry sample of sand was sheared in a direct shear box where it failed under a shear stress of 10 lb/in.² when the normal stress was 14.2 lb/in.² What is the angle of shearing resistance (angle of internal friction) of this sandy soil?

Angle of shearing resistance $\phi = 35°$ *Answer*

8.39   For the conditions in Problem 8.38, draw a Mohr circle of failure for the sample and determine the values of major and minor principal stresses within the sample at time of failure. Also, on the same circle, work the positions of the major and minor principal planes.

$\sigma_3^p = 9.1$ lb/in.², $\sigma_1^p = 33.6$ lb/in.²   *Answer*

The inclination of the major principal plane is 62°.
The inclination of the minor principal plane is 28°.

TABLE P-8.40

	Soil A				Soil B				Soil C		
Test	1	2	3	4	1	2	3	4	1	2	3
Horizontal shear force, lb	21.9	22.7	24.0	28.0	7.1	10.5	17.3	24.2	10.1	10.4	10.5
Vertical load, lb	7.0	10.0	15.0	30.0	12.0	20.0	30.0	42.0	20.0	25.0	30.0

8.40   Table P-8.40 shows thows the results obtained in series of tests conducted on three different samples of soil in a direct shear box. The area of each sample in the shear box was 4 in.² Plot these results and determine the values of apparent cohesion and internal friction for each soil. Judging from the shape of each relationship, what is the probable type of soil in each sample?

Soil A, $c = 5$ lb/in.², $\phi = 15°$. (Soil has a high value of angle of internal friction   *Answer*
and a reasonable value of cohesion. Thus it is probably a sandy-clay soil.)
Soil B, $c = 0$ lb/in.², $\phi = 30°$. (Soil has a high value of angle of internal friction and no cohesion. Thus it is probably a sandy soil.)
Soil C, $c = 10$ lb/in.², $\phi = 1°$. (Soil has almost no angle of internal friction and a high value of cohesion. Thus it is probably a clay or silty-clay soil.)

8.41   A sample of cohesionless soil was tested in a drained shear box test. At failure, the shear stress was 49.7 lb/in.², under a normal stress of 160.0 lb/in.² Determine, by the use of a Mohr circle and then analytically, the values of the angle of internal friction of the soil, the resultant stress, and the angle of inclination of the plane of principal stress to the failure plane.

Answer $\phi = 17°15'$; resultant stress $\sigma_r^m = 168$ lb/in.², $\omega = 53°37'$

8.42 A series of shear tests was conducted on a soil sample. The values of major and minor principal stresses at failure for each sample are recorded in Table P-8.42. Determine the angle of internal friction and cohesion of this soil.

Answer Plot the Mohr circles for the given conditions and construct the best common tangent to these circles (strength envelope). From this envelope, $\phi = 30°$ and $c=0$.

8.43 Explain the terms total pressure, effective pressure, and pore-water pressure as applied to saturated soils. Then show how these pressures are related to each other. Also, explain the influence of pore-water pressure on the shear strength of soils.

Answer Refer to Chapter 4 and to Chapter 8, Section 8.4.

8.44 The values in Table P-8.44 were obtained from a series of undrained triaxial tests on samples of a saturated soil. Determine the values of cohesion and angle of internal friction for this soil in terms of (a) total stress and (b) effective stress.

TABLE P-8.42

Test Number	σ_3^p (lb/in.²)	σ_1^p (lb/in.²)
1	12.0	36.5
2	17.5	52.5
3	25.0	75.0

TABLE P-8.44

Lateral pressure σ_3^p (lb/in.²)	20	36	49
Pore-water pressure u_w (lb/in.²)	5	10	14
Principal stress difference at failure $\sigma_d = \sigma_1^p - \sigma_3^p$ (lb/in.²)	55	72	86

Answer (a) $\phi_u = 20°$, $c_u = 12$ lb/in.²
(b) $\phi_1 = 27°$, $c_1 = 10$ lb/in.²

8.45 Table P-8.45 shows the readings obtained during a series of consolidated-undrained tests on an undisturbed sample of overconsolidated clay. Plot the failure envelope for the soil in relation to (a) total stress and (b) effective stress. Comment on the results. From the plotting of the failure envelopes, determine the apparent and effective angles of internal friction and cohesion.

(a) $\phi_u = 11°$, $c_u = 7.5 \, \text{lb/in.}^2$ *Answer*
(b) $\phi_1 = 14°$, $c_1 = 6.0 \, \text{lb/in.}^2$

8.46 An unconfined compressive strength test was conducted on a specimen of clay soil 2 in. in diameter and 4 in. in height. The specimen was found to have shortened by 0.25 in. at failure under an axial load of 120 lb. Determine the unconfined compressive strength and the shear strength of this soil.

TABLE P-8.45

Cell Pressure (lb/in.²)	Deviator Stress at Failure (lb/in.²)	Pore-Water Pressure at Failure (lb/in.²)
5.0	26.0	15.0
10.0	30.0	14.0
30.0	40.0	10.0
47.0	46.0	6.0

Area of sample at failure is   *Answer*

$$A_2 = \left[\frac{A_1}{1 - (\Delta L/L_1)} \right] = \left[\frac{3.14}{1 - (0.25/4.0)} \right] = 3.35 \, \text{in.}^2$$

Unconfined compressive strength

$$q_u = \frac{\sigma_1^p}{A_2} = \frac{120}{3.35} = 35.82 \, \text{lb/in.}^2$$

Shear strength

$$S = c_u = \frac{q_u}{2} = 17.91 \, \text{lb/in.}^2$$

stability
analysis
9

9.1 INTRODUCTION

Conditions of Performance

The *function of the structure* above or within the soil system is to carry loads and resist stresses. In order to perform this function, the loaded structure and the supporting soil system must be in a state of *stable static equilibrium* and each of the components must be strong enough to make this performance possible.

For such a state to exist, three sets of conditions must be fulfilled simultaneously:

1. the *conditions of static equilibrium* based on Newton's principles.
2. the *conditions of stability* based on the concept of incipient load or displacement.
3. the *conditions of strength* based on the concept of critical (ultimate) stress.

Since the structure is assumed a priori to satisfy these conditions, the study in this chapter is restricted only to the fulfillment of these conditions in the soil system.

Stable Equilibrium

A loaded soil system is in a state of stable equilibrium if, for any incipient small increment in load, a resistance is developed. If the small increment in load causes soil rupture, rotation, and or slide, the system becomes unstable and a partial or complete failure may follow.

Modes of Failure

Among many possible foundation failures, the *break-in failure* (Fig. 9-1a), the *rotation failure* (Fig. 9-1b), and the *slide failure* (Fig. 9-1c) are most typical and are investigated in this chapter.

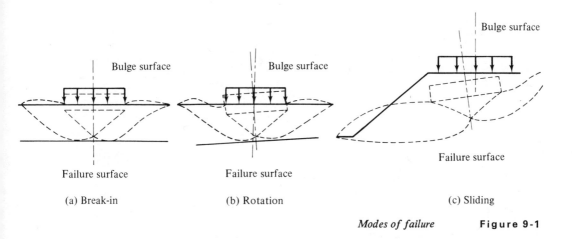

| (a) Break-in | (b) Rotation | (c) Sliding |

Modes of failure   **Figure 9-1**

9.2 *FORCE INTERACTION*

Basic Terms

The stability analysis involves the comparison of the two systems of forces: (a) the acting forces and (b) the resisting forces, both acting along the *potential surface of failure* (rupture).

The *acting forces* are those forces developed by causes introduced and analytically defined in the preceding chapters (applied loads, soil moisture, and volume change).

The *resisting forces* are essentially the shearing forces acting tangentially on the surface of failure. The ultimate values of these forces are influenced by many factors; yet, for practical purposes, they may be approximated by the strength formulas developed earlier in Sections 8.3, 8.4, and 8.5 and summarized in Table 9-1 of this section.

TABLE 9-1　　*Resisting Shearing Stress*

Soil Type	Expression	Equation Number
Cohesionless sands $\phi > 0$	$s = (p - u_w) \tan \phi$	(8-21)
Overconsolidated clays $\phi > 0$	$s = c + (p - u_w) \tan \phi$	(8-23)
Saturated clays $\phi = 0$	$s = c = \dfrac{p_1 - p_3}{2}$	(8-24)

SYMBOLS:　s = resisting shearing stress, p = normal pressure, u_w = hydrostatic pressure, c = cohesion, p_1, p_3 = principal normal pressure, ϕ = angle of internal friction.

Governing Equation

The condition of failure, already given by (8-19), is here expressed in terms of the flow value

$$N_\phi = \tan^2\left(\frac{\pi}{4} + \frac{\phi}{2}\right) \tag{9-1}$$

as

$$p_1 = 2c\sqrt{N_\phi} + p_3 N_\phi \tag{9-2}$$

Special Forms

This equation takes two special forms under the following conditions.
For $c = 0$,

$$p_1 = p_3 N_\phi \tag{9-3}$$

For $\phi = 0$,

$$p_1 = 2c + p_3 \tag{9-4}$$

The application of these equations is shown in Problems 9.1 and 9.2.

9.3 BEARING STABILITY

Ultimate Bearing Capacity

The determination of the bearing capacity of a soil system is the most important problem of engineering soil mechanics. When a load is applied on a limited area of the surface, the surface deforms (settles). The relation between the bearing stress (pressure) and the associated settlement can be represented by a diagram, called the *load-settlement* curve, similar to the stress-strain curve of other materials. These curves take different shapes, depending on the type of soil involved. The *ultimate bearing capacity* (stress), defined as the stress at failure, is well defined for dense sands and insensitive clays, whereas it is progressive and not well defined for the loose sands and sensitive clays.

Presumptive Bearing Capacity

The oldest and least-dependable method of determining the bearing capacity of the soil system is based on past experiences incorporated in state and municipal codes in the form of tables of presumptive bearing pressures (allowable pressures). These tables are a useful guide to local practice; the data, which are usually based on descriptive classifications, are poor approximations at best.

Tested Bearing Capacity

A more dependabable, yet often misleading, method involves the determination of the bearing capacity by the plate load test at the given site. The testing equipment, procedure, and calculations related to this method are discussed in Problems 9.3 through 9.8.

Prandtl Method

The analytical methods for the prediction of the ultimate bearing capacity of soil systems stem from Prandtl's *plastic equilibrium theory*, developed originally for the analysis of failure in a block of metal under a long, narrow strip loading.

According to Prandtl, five zones exist in the material at failure (Fig. 9-2).

1. A *wedge-shaped zone* under the loaded area, pressing the material downward as a unit $(CC'B)$.
2. Two *zones of all radial failure plans* through A and B, each bounded by a logarithmic spiral $(CBD, C'D'B)$.
3. Two *triangular zones* forced by pressure upward and outward as two independent units $(CDE, C'D'E')$.

Although the behavior of a soil system is not in close agreement with Prandtl's model, the mechanism of failure of both systems is similar and allows the utilization of Prantl's ultimate stress equation for the calculation of the ultimate bearing capacity of cohesive soil of known c, ϕ under the base of a narrow strip footing.

With these restrictions, the ultimate bearing capacity per unit area is

$$q_u = c \cot \phi (e^{\pi \tan \phi} N_\phi - 1) \tag{9-5}$$

where N_ϕ is given by (9-1) and c, ϕ are those of Table 9-1.

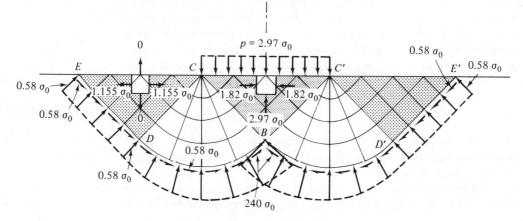

Figure 9-2 *Slip lines and forces in plastic region, Prandtl's model*

(O. Hoffman and G. Sachs, *Introduction to the Theory of Plasticity for Engineers*, McGraw-Hill, New York, 1953, p. 133.)

If $\phi = 0°$, the logarithmic spiral reduces to a circular arc, and

$$q_u = (2 + \pi)c = (5.14159\ldots)c$$

The derivation of (9-5) and its applications are given in Problems 9.9, 9.10, and 9.11.

Terzaghi Method

The solution (9-5) advanced by Prandtl (1921) is, of course, only a particular solution of this problem, in which the width of the strip and its position below the ground surface are neglected and γ is assumed to be zero.

Although efforts were made by others to present a more complete solution, it was Terzaghi (1943) who developed the first rational and practical approach to this problem. His method includes three dominant factors: (a) the weight of soil, (b) the effect of surcharge, and (c) the strength parameters of the soil, and is more general than any other previously proposed method.

This solution recognizes two distinct cases:

1. the *general shear*, the load-settlement curve of which is designated by C_1 in Fig. 9-3a.
2. the *local shear*, the load-settlement curve of which is designated by C_2 in Fig. 9-3b.

The approximate value of the bearing capacity of long footing at or below the ground surface for any soil is given below by two expressions, each corresponding to one of these cases.

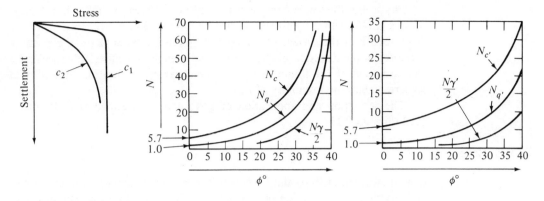

(a) Stress-settlement curves   (b) Case (I) — charts   (c) Case (II) — charts

Coefficients of ultimate bearing capacity,   **Figure 9-3**
Terzaghi method

(D. W. Taylor, *Fundamentals of Soil Mechanics*,
Wiley, New York, 1948, p. 576. Copyright
1948 by Donald W. Taylor. Reprinted by
permission of John Wiley & Sons, Inc.)

For Case 1:

$$q_u = cN_c + \tfrac{1}{2}\gamma bN_\gamma + \gamma dN_q \qquad\qquad (9\text{-}6)$$

For Case 2:

$$q_u = \tfrac{2}{3}cN'_c + \tfrac{1}{2}\gamma bN'_\gamma + \gamma dN'_q \qquad\qquad (9\text{-}7)$$

In these equations, N_c, N_γ, N_q and N'_c, N'_γ, N'_q are dimensionless coefficients given by charts in Fig. 9-3b and c, respectively; b is the width of the footing, d is the depth, γ is the unit weight of soil, and c is the cohesion.

The results obtained by (9-6) and (9-7) are approximate, but if compared with more recent and more sophisticated solutions of G. G. Meyerhof (1951) and V. V. Sokolovski (1960), they prove to be on the safe side.

Ultimate Bearing Capacity of   **TABLE 9-2**
Circular and Rectangular Footings

Case	Circular	Rectangular
1	$q_u = 1.3cN_c + 0.3\gamma bN_\gamma + \gamma dN_q$	$q_u = 1.3cN_c + 0.4\gamma bN_\gamma + \gamma dN_q$
2	$q_u = 1.3cN'_c + 0.3\gamma bN'_\gamma + \gamma dN'_q$	$q_u = 1.3cN'_c + 0.4\gamma bN'_2 + \gamma dN'_q$

NOTE: Symbol meanings are given in equations (9-6) and (9-7). The numerical values of N_c, N_γ, N_q and N'_c, N'_γ, N'_q must be taken from the charts of Fig.9-3b and c.

Physically, the soil system begins to fail in the break-in move but, finally, always fails by tilting. This result must be interpreted as a consequence of the imperfection of the surface of the ground. As tilting takes place, the pressure increases in the direction of the overturning, which, of course, only accelerates the formation of the failure mode. To some extent, the bearing instability of soils resembles the instability of columns.

The derivation and application of (9-6) and (9-7) are given in Problems 9.12 through 9.17.

Special Cases

The application of (9-6) and (9-7) is restricted to the solution of long, narrow strip footings. The solution of this problem for other shapes is very involved; generally numerical methods or approximate formulas are employed.

Two special shapes (circular and rectangular footings) are considered here and solved for the general shear (Case 1) and the local shear (Case 2) in Table 9-2 (Problems 9.10, 9.11).

9.4 SLOPE STABILITY

General

The failure of a soil mass located beneath a soil slope is called a *landslide*, the effects of which can be spectacular and in some instances disastrous. The occurrence of landslides may be *progressive* or *instantaneous* without warning, inflicted by human action (excavation, undercutting, overloading, etc.) or by natural causes within, on the top, or below the slope (disintegration, decomposition, drainage, seepage, frost, etc.).

Modes of Failure

In general, the slope failure takes on one of the following modes (shapes):

1. The *base failure* (Fig. 9-4a) in cohesive soils is preceded by the formation of tension cracks above the upper edge of the slope, followed by a shear failure along a slip surface (separation) that intersects the ground at some distance from the base of the slope.
2. The *slope failure* (Fig. 9-4b) in cohesive soils is again preceded by the formation of tension cracks at the top, followed by a shear failure along a slip surface that intersects the slope.
3. The *toe failure* (Fig. 9-4c) in soils of $\phi > 0$ forming steep slopes is a shear failure along a slip surface that intersects the toe of the slope.

Although this group classification is a convenient descriptive approach, sometimes the actual failures occur in a less distinct form as a combination of two modes.

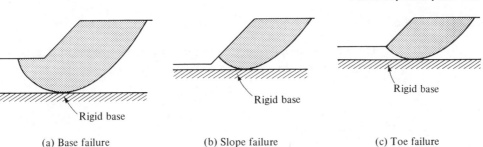

(a) Base failure   (b) Slope failure   (c) Toe failure

Modes of slope failure   **Figure 9-4**

Analysis of Failure

The analytical prediction of the slope failure in a given system requires the location of the slip (failure, rupture) surface, the determination of the acting forces, and the prediction of the available reactive forces on it.

The safety factor of a given slip surface is then

$$F_s = \frac{\text{forces opposing slip (ultimate shear)}}{\text{forces causing slip (working shear)}} \tag{9-8}$$

where the *ultimate shear* is the sum of the critical shearing stresses (Table 9-1) acting along the slip surface and the *working shear* is the sum of the shearing stresses produced by causes (deadload, water, etc.) along the same surface.

The stability analysis is largely a trial-and-error process, accomplished by numerical or graphical methods. Because of the cyclic (repetitive) nature, the calculations and the graphical constructions are well suited to computer programming, which, of course, reduces the involved labor considerably.

Numerous analytical methods have been developed throughout the years and occur in the literature under various and sometimes conflicting names. The most common methods in this group are

1. K. Culmann method (1866)
2. K. E. Peterson method (1915)
3. D. W Taylor method (1936)
4. W. Fellenius method (1935)

All involve the *principle of limit equilibrium* of tangential forces on the slip surface and all possess two deficiencies:

1. The problem is statically indeterminate and cannot be solved without the deformation conditions.
2. The parameters of strength (c, ϕ) and the pore-water pressure u_w must be estimated; in actual slopes, a great uncertainty exists in this respect.

Thus considerable judgment and great care must be exercised in evaluating the analytical results, which again are only approximations at best.

Culmann Method

The basic assumption of this method is the occurrence of a slip plane (not curved surface) through the toe of the slope (Fig. 9-5). The limiting equilib-

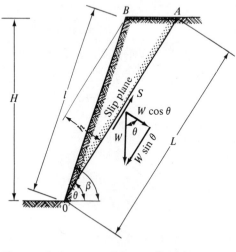

Figure 9-5   *Culmann slip plane*

rium along this plane gives

$$\underbrace{F_s W \sin \phi}_{F_s T} = \underbrace{W \cos \theta \tan \phi + cL}_{S} \tag{9-9}$$

where W = total weight of the soil above $\overline{OA}$
θ = slope of the slip plane of length L
c, ϕ = parameters of shearing stress (Table 9-1)
T = resultant working forces
S = resultant opposing forces
F_s = safety factor

The total weight of the soil mass above the slip plane is

$$W = V\gamma = \frac{H^2 \gamma}{2} \frac{\sin (\beta - \theta)}{\sin \beta \sin \theta} \tag{9-10}$$

and in terms of (9-9) leads to the expression of the stability factor,

$$N_s = \frac{\gamma H}{c} = \frac{2 \sin \beta}{\sin (\beta - \theta)(\sin \theta - \cos \theta \tan \phi)} F_s \tag{9-11}$$

where H = slope height
β = incidence angle (Fig. 9-5)
$\gamma = \gamma_m$ = density of soil

The critical values of θ and H, computed for $F_s = 1$ from $\partial N_s / \partial \theta = 0$, are then

$$\theta_{\text{CR}} = \frac{\beta + \phi}{2}, \qquad H_{\text{CR}} = \frac{2c \sin \beta \cos \phi}{\gamma \sin^2 [(\beta - \phi)/2]} \qquad (9\text{-}12)$$

and for $\beta = \pi/2$ become

$$\theta_{\text{CR}} = \frac{\pi}{4} + \frac{\phi}{2} \qquad H_{\text{CR}} = 4\left(\frac{c}{\gamma}\right) \tan\left(\frac{\pi}{4} + \frac{\phi}{2}\right)$$

The Culmann method gives good results for vertical or almost vertical slopes ($\beta \cong 90°$) and *intolerable errors* on the danger side *for flat slopes*. The derivation and applications of (9-12) are given in Problems 9.18, 9.19, and 9.20.

Petterson Method

The first general method for the slope stability, known as the *circular arc method*, has been developed by Petterson and is frequently called the *Swedish method* or the Petterson method.

The assumptions of this method are that (a) the shear strength of the soil system is known, (b) the soil system rests on a rigid base which is not penetrated by the slide surface, (c) the soil slope is the plane between two horizontal planes, and (d) the slide (failure, rupture) surface is cylindrical with a normal axis to the plane of the slope cross section.

For the stability determination of a particular slope, it is necessary to determine the radius and the center of the circle along which the slope will eventually fail.

For the circular surface of Fig. 9-6, the limiting equilibrium equation (9-9), in the absence of normal pressure along the *slip curve*, is

$$F_s T = c \int_0^A dL = cL \qquad (9\text{-}13)$$

and the safety factor is

$$F_s = \frac{cL}{T} \qquad (9\text{-}14)$$

where $L = \overline{OA}$ and T is the sum of the tangential working forces along L.
The value of F_s varies with the geometry of the slip arc (its radius and

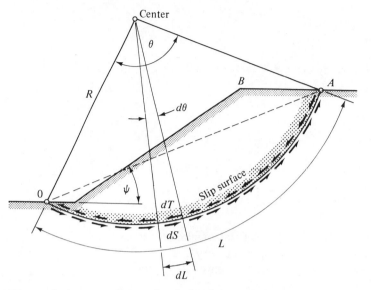

Figure 9-6   *Petterson's cylindrical slip surface*

center), and discovering the circle that corresponds to the minimum value of F_s is a formidable task (Problems 9.21 to 9.26).

For a stable slope, the minimum F_s should generally be equal to or greater than 1.5.

Taylor Method

A general solution based on the *friction-circle method* was developed by Taylor and is explained in Problems 9.34 and 9.35. Tables and charts prepared by Taylor, Fellenius, and others are given with applications in Problems 9.29 to 9.33. Although these charts are valid only for simple, homogeneous, finite slopes under undrained conditions, they may also be used as rough estimates for other, more complex cases.

Fellenius Method

If the Petterson method is the curvilinear generalization of the Culmann method, neglecting the normal pressure on the slip line, the Fellenius method is a complete generalization of the Culmann method for a curved slide.

For slopes with irregular surface and consisting of material—the parameters c, ϕ of which vary with depth—the Fellenius method can be conveniently formulated as the *finite slice method* and the solution executed in tabular form (Problem 9.16) or by means of charts (Problems 9.27 and 9.28).

Modes of Failure—Force Interaction

9.1 A series of triaxial tests were performed on a soil sample that has a cohesion $c = 1.35\ t/ft^2$ and an angle of internal friction $\phi = 24°$. Determine the vertical pressure that would be required to produce failure for a mass of this soil at a confining pressure $p_3 = 0.5\ t/ft^2$.

Assuming that equation (9-2) is valid in this case,   *Solution*

$$p_1 = 2c\sqrt{N_\phi} + p_3 N_\phi \tag{a}$$

where $\sqrt{N_\phi} = \tan\left(45° + \dfrac{\phi}{2}\right) = \tan\left(45 + \dfrac{24}{2}\right) = 1.539$

$$N_\phi = (1.535)^2 = 2.372$$
$$p_3 = 0.5\ t/ft^2$$
$$c = 1.35\ t/ft^2$$

Then

$$p_1 = (2)(1.35)(1.539) + (0.5)(2.372) = 5.346\ t/ft^2$$

9.2 A sample of uniform sand is tested in a drained triaxial test, with a lateral confining pressure of 21 lb/in². Compute the vertical pressure at which the sample fails.

Since the material tested is sand (cohesionless), the value of cohesion $c = 0$; and *Solution*
(9-3) is used to determine the value of the vertical pressure p_1 at failure, where

$$p_1 = p_3 N_\phi \tag{a}$$

and

$$p_3 = 21\ lb/in^2.$$

From Table 8.3, the angle of internal friction ϕ for a uniformly dense sand is assumed to be about 35 degrees and

$$N_\phi = \tan^2\left(45 + \frac{\phi}{2}\right) = \tan^2\left(45 + \frac{35}{2}\right) = (1.921)^2 = 3.68$$

Substituting the preceding quantities in (9-3) yields

$$p_1 = p_3 N_\phi = 21(3.68) = 77.3\ lb/in^2.$$

Plate-Loading Test

9.3 Describe the plate-loading (plate-bearing) test.   253

Answer The function of the plate-loading test is to measure the soil's bearing capacity by applying a compressive stress to the soil surface through a rigid bearing area and measuring the deflection under various load increments. It is a field test carried out on the natural soil with large-scale apparatus and a test area of several square feet and thus with the soil being loaded in much the same manner during the test as in practice under actual footings.

The apparatus used in the plate-loading test consists essentially of the following elements: (a) a rigid loading plate, (b) equipment to apply a load to the plate, and (c) gages to measure that load and the resulting deformation of the plate (which is assumed to be the same as the deformation of the soil mass). The standard test consists of a circular plate of 30 in. diameter, which is stiffened with plates of 26 in. and 22 in. diameter placed on top of it. A schematic diagram of the apparatus is shown in Fig. P-9.3. The load is applied by a dead load of a heavy truck or trailer,

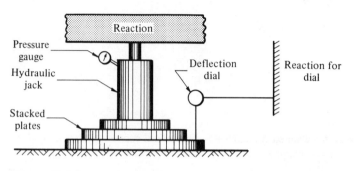

Figure P-9.3

which is then transmitted to the plate by a hydraulic jack in which a pressure gage is connected to the output end to measure the load. An independent datum for the measurement of the settlement of the plate is provided by a frame whose supports rest on the ground at points unaffected by the settlement of the plate or the loading vehicle. The settlement of the plate is considered as the average of the readings of a number of dial gages arranged symmetrically around the periphery of the plate.

9.4   Describe the test procedure for conducting the plate-loading test.

Answer The test procedure is outlined in the following steps:

1.   The conditions of the moisture content and dry density of the soil in the test area are brought to conditions similar to those that are likely to exist when the soil mass has reached a state of equilibrium subsequent to the construction of the foundation.

2.   Great care should be taken to seat the plate accurately, by carefully leveling the test area. On coarse-grained soils that are difficult to level accurately, the plate can be seated on a layer of fine dry sand, which should not be thicker than $\frac{1}{4}$ in. at any place.

3.   In loading the plate, two methods are used—namely, the P. C. A. method and

the Corps of Engineers method. The selection between either method depends on the purpose for which the test is performed.

(a) The P. C. A. method: The plate is first seated by applying a load equivalent to a pressure of 1 lb/in.^{2} and then releasing it after a few seconds. A load sufficient to cause approximately a 0.01-in. settlement is next applied; and when the rate of settlement is less than 0.001 in./min, the average of the readings of the dial gages is noted. The load as measured by the pressure gage attached to the jack is also noted. The load is then increased in increments causing approximately 0.01-in. deformation each, and the load and settlement are both recorded. The procedure is repeated until a total settlement of not less than 0.09 in. has been produced.

(b) The Corps of Engineers method: The plate is seated in the same manner as in the P. C. A. method but the loading test is restricted to the application of only one load increment. A load equivalent to 10 lb/in.^{2} for the 30-in.-diameter plate is applied in 10 seconds and held until there is no increase in settlement or until the rate of settlement is less than 0.002 in./min, when the dial gages for recording the settlement are read.

9.5 Show how the results from a typical plate-loading test are expressed. Also, show how the results are used to indicate the bearing capacity of the soil.

Answer

The average pressure under the loading plate is usually plotted against the corresponding values of average settlement, as shown in Fig. P-9.5. The resulting curve is normally convexing upward and has no straight segments. A factor known as the modulus of soil reaction k is determined from the results of this test and is considered as an indication of the bearing capacity of the soil mass at the location of the test. The bearing capacity of the soil is directly proportional to k.

The calculation of k from test results depends on the method used for loading the plate (the P. C. A. or the Corps of Engineers).

1. If the P. C. A. method was used in loading the plate, k is taken as the slope of the line passing through the origin and the point on the curve corresponding to 0.05-in. settlement.

$$k = \frac{p}{0.05} \text{ lb/in.}^2/\text{in.}$$

where p = the pressure in lb/in.^{2} to cause 0.05-in settlement.

2. If the Corps of Engineers method was used in loading the plate, only the settlement corresponding to a pressure of 10 lb/in.^{2} is determined, and thus no curve between pressure and settlement is plotted. In this case, k is simply expressed as

$$k = \frac{10}{d} \text{ lb/(in.}^2)/(\text{in.})$$

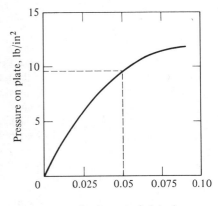

Figure P-9.5

where d = settlement in inches corresponding to a pressure of 10 lb/in.^{2}
The two procedures will only give identical results when k = 200 lb/in.^{3}

9.6 Are there any corrections to be made on the value of the modulus of soil
reaction k after it is determined from the results of the plate-loading test? Explain
the type of correction and the procedure of introducing it into the determined value
of k.

Answer There are two basic corrections to be made on the computed value of k from the
results of the field plate-loading test.

1. Correcting the field value of k to the most unfavorable subgrade soil that can
   be expected. Since it is usually impractical to conduct tests in the field on representative soils at the various densities and water contents that will approximate the service conditions to be expected under the foundation, it is necessary
   to adjust the field value of k to the most unfavorable subgrade condition that
   can be expected. This step is accomplished by running two series of consolidation tests in the laboratory on representative samples of the soil mass—one on
   the soil in its natural condition of moisture content and dry density and the
   other after the soil has been soaked under a surcharge of 5 lb/in.^{2} The results
   of these tests are plotted as pressure against final deformation when consolidation is complete, as shown in Fig. P-9.6a and b. The correction of the field
   value of k depends on the method used for its determination.
   (a) If the P. C. A. method was used in determining the field value of k
        (Problem 9.4), the corrected value of k is k_s and is expressed as

   $$k_s = k\left(\frac{p_s}{p}\right) \text{ lb/in.}^3$$

   where p is the pressure required in the plate-loading test to cause a settlement of 0.05 in. and p_s is the pressure required in a consolidation test
   on the soaked specimen to cause a deformation equal to that produced
   by pressure p in the consolidation test on the unsoaked specimen (Fig.
   P-9.6a).
   (b) If the Corps of Engineers method was used in determining the field value
        of k (Problem 9.4), then

   $$k_s = k\left(\frac{d}{d_s}\right)$$

   where k_s is the corrected value of the field modulus of soil reaction k, d
   is the compression resulting from a consolidation test on a natural sample
   of the soil under a pressure p of 10 lb/in.^{2}, and d_s is the compression
   produced in a consolidation test on the soaked specimen under a pressure
   p = 10 lb/in.^{2}. (Fig. P-9.6b).
   It is assumed, in this case, that the ratio of the deflections in the unsaturated
   and saturated consolidation tests is approximately the same as the ratio of the
   deflections in the field on unsaturated and saturated soil masses.
2. The second correction is that resulting from the deflection (bending) of the
   plate when testing materials with a high modulus of subgrade reaction. When

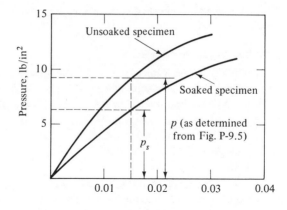

(a)

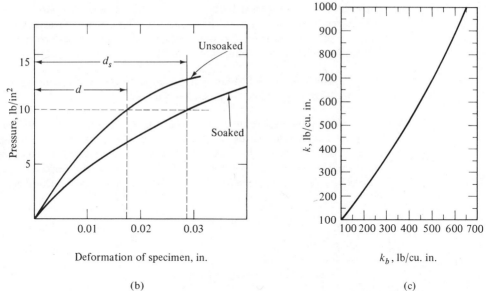

Deformation of specimen, in.

k_b, lb/cu. in.

(b)

(c)

(Corps of Engineers, *Engineering and Design,*   **Figure P-9.6**
Rigid Pavements, EM 1110-45-303, 1958, p. 40.)

deflection of the plate takes place, the volume of soil displaced under the plate
will thus be different from that determined by dials on the circumference of
the plate. These values can be corrected for bending by using the data in Fig.
P-9.6c, in which k is the value of the modulus of subgrade reaction produced
from the field test and k_b is the value of the modulus corrected for plate
bending.

9.7   List the chief sources of error in using the plate-loading test data in designing actual foundations.

Answer   The chief sources of error are

1. The size of the plate has an especially important effect on the k value determined from the results of the field plate-loading test; and in any statement concerning the value of the modulus, the size of the plate must be indicated.
2. The difference in area, and possibly in shape, between the loaded plate and that of the actual foundation.
3. The greater depth of soil affected by the structure as compared with the test plate.
4. The deformation of the soil mass under the loaded plate is a combination of both elastic and plastic deformations of unknown proportions.
5. The nonuniformity and nonhomogenity of the soil mass, which tends to complicate the interpretation of the test results.
6. The short duration of the test, which does not allow for the full development of the plastic deformation of the cohesive soils.

In spite of the preceding limitations, a plate-loading test, when properly conducted and correctly interpreted, is valuable for predicting the settlement and the bearing capacity of cohesionless soils, as well as some varved and fissured clays that are difficult to sample and test in the undisturbed state.

9.8   Explain how the safe bearing capacity of a soil mass can be approximately determined from the results of a plate-loading test.

Solution   After plotting the data from the plate-loading test (pressure versus settlement), the value of the ultimate bearing capacity of the plate is defined either with respect to a failure point represented by a definite break in the load-settlement curve, or by the tangential deviation point (curves I and II, Fig. P-9.8), or with respect to an arbitrary selected value of settlement if the curve does not show any breaking point (curve III).

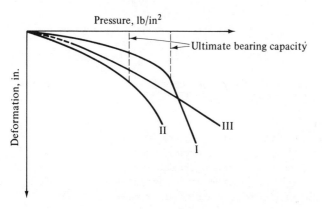

Figure P-9.8

This ultimate bearing capacity of the soil under the plate, when divided by a factor of safety of $F_s = 2$, gives the safe bearing capacity of the plate.

Prandtl Method

9.9 Discuss the origin of the Prandtl equation (9-5), which is used in calculating the ultimate bearing capacity of soil.

Solution

The derivation of (9-5) requires a knowledge of the fundamentals of the mathematical theory of plasticity of solids* and is beyond the level of this book. For this reason, the Prandtl equation is traditionally omitted or given without proof with reference to his original work.† The reprint of this paper is also included in the more recently published collection of Prandtl's papers.‡

9.10 Determine the bearing capacity of a foundation supporting soil that has the following properties: angle of internal friction $\phi = 10°$, cohesion $c = 200 \, \text{lb}/\text{ft}^2$. Assume that the foundation soil is weightless.

Solution

In this case, where the foundation soil is assumed to be weightless, the bearing capacity is determined by using the Prandtl equation (9-5),

$$q_u = c \cot \phi (e^{\pi \tan \phi} N_\phi - 1) \tag{a}$$

where

$$c \cot \phi = (200)(\cot 10°) = (200)(5.67) = 1134.2 \, \text{lb}/\text{ft}^2$$

$$e^{\pi \tan \phi} = e^{(3.14) \tan 10°} = e^{(3.14)(0.1763)} = 1.740.$$

$$N_\phi \text{ (from 9.1)} = \tan^2 \left(\frac{\pi}{4} + \frac{\phi}{2} \right) = \tan^2 \left(45° + \frac{10°}{2} \right)$$

$$= \tan^2 50° = (1.192)^2 = 1.421$$

Thus the ultimate bearing capacity of the soil is

$$q_u = (1134.2)[(1.740)(1.421) - 1] = (1134)(1.473) = 1672 \, \text{lb}/\text{ft}^2$$

9.11 Solve Problem 9.9, assuming that the foundation soil is a purely cohesive clay with cohesion $c = 400 \, \text{lb}/\text{ft}^2$.

Solution

For a purely cohesive soil, the angle of internal friction $\phi = 0$, and the ultimate bearing capacity q_u computed by the Prandtl equation (9-5) is

$$q_u \cong 5.14c \cong 5.14(400) = 2056 \, \text{lb}/\text{ft}^2$$

*O. Hoffman and G. Sachs, *Introduction to the Theory of Plasticity for Engineers*, Mc-Graw-Hill, New York, 1953, pp. 113–134.
†L. Prandtl, Über die Eindringung Festigkeit (Härte) Plastischer Baustoffe und die Festigkeit von Schneiden, *Zeitschrift für Angewandte Mathematik und Mechanik*, Vol. 1, 1921, pp. 15–20.
‡W. Tollmien, H. Schlichting, and H. Goertler, Ludwig Prandtl, Gesammelte Abhandlungen Zur Angewandten Mechanik, *Hydro und Aerodynamik*, Vol. I, Springer, Berlin, 1961, pp. 104–112.

Terzaghi Method

9.12 Discuss the origin of the Terzaghi equations (9-6) and (9-7), which are used in calculating the ultimate bearing capacity of soil.

Solution The derivation of equations (9-6) and (9-7) is based on a series of assumptions that require considerable discussion and cannot be presented here. For this reason, Terzaghi equations are traditionally given without proof with reference to his original work* and to the work of Meyerhof.†

9.13 Describe the limiting conditions for which general or local shear failure should be assumed to prevail at a given site when Terzaghi's ultimate bearing capacity equations (9-6) and (9-7) are used.

Answer It is difficult to define precisely the limiting conditions for which general or local shear failure would prevail. However, the following guidelines developed by Peck and Singh (1959) can be used:

1. If the failure of a soil possessing both cohesion and internal friction ($c - \phi$ soil) occurs at a small strain in a shear test (e.g., less than 5 percent), failure would probably occur in the field by general shear.
2. If, on the other hand, a $c - \phi$ soil fails in a shear test at strains of over 10 percent, local shear failure in the field would seem more probable.
3. In a fairly soft, or loose and compressible soil that would undergo large deformations under the foundation before the failure zone develops, failure by local shear is most probable.
4. For cohesionless soils, if the angle of internal friction ϕ is more than 36 degrees, general shear failure is probable; and when ϕ is less than 29 degrees, local shear failure may be assumed. For intermediate values of ϕ, between 29 and 36 degrees, the values of bearing capacity factors are obtained by interpolation between the limiting values for local and general shear failure conditions.

9.14 Consider a strip footing 6 ft wide and built at a depth of 5 ft in a soil that has the following characteristics: density $\gamma = \gamma_m = 110 \, \text{lb/ft}^3$, cohesion $c = 500 \, \text{lb/ft}^2$, and angle of internal friction $\phi = 25°$. Determine the allowable bearing capacity of the foundation, assuming a safety factor of 2.0.

Solution The soil does not appear to be soft and loose (has high values of c and ϕ). Thus the ultimate bearing capacity can be determined by using Terzaghi's equation (9-6) for general shear failure.

$$q_u = cN_c + \tfrac{1}{2}\gamma_m bN_\gamma + \gamma_m dN_q \tag{a}$$

In equation (a), values of c, γ_m, b, and d are given in the problem statement. Values of the dimensionless parameters N_c, N_γ, and N_q are determined from Fig. 9-3b. Thus for $\phi = 25°$,

*K. Terzaghi, *Theoretical Soil Mechanics*, Wiley, New York, 1943, pp. 118–136.
†G. G. Meyerhof, "The Ultimate Bearing Capacity of Foundations," *Geotechnic*, Vols. 1, 2, 1951, pp. 301–332.

$$N_c = 25, \quad N_q = 13, \quad N_\gamma = 10 \tag{b}$$

Substituting values of these dimensionless parameters in (9-6), which is restated above in equation (a), yields the ultimate bearing capacity

$$q_u = (500)(25) + \tfrac{1}{2}(110)(6)(10) + (110)(5)(13)$$

$$= 12500 + 3300 + 7150 = 22950 \text{ lb/ft}^2 = 11.475 \ t/\text{ft}^2 \tag{c}$$

The allowable bearing capacity is

$$q_{allow} = \frac{q_u}{F_s} = \frac{11.475}{2} \cong 5.73 \ t/\text{ft}^2 \tag{d}$$

9.15 Solve Problem 9.13, assuming the value of the angle of internal friction of the soil to be 30 degrees. Compare the values of the ultimate bearing capacity q_u calculated for $\phi = 30°$ and that in Problem 9.12 for $\phi = 25°$, and comment on the results.

Solution

As in Problem 9.14, q_u is given by (9-6) as

$$q_u = cN_c + \tfrac{1}{2}\gamma_m bN_\gamma + \gamma_m dN_q$$

where c, γ_m, b, and d are the same as before. Values of the dimensionless parameters N_c, N_γ, and N_q are determined from Fig. 9.3b for $\phi = 30°$ as

$$N_c = 37, \quad N_q = 22, \quad N_\gamma = 22$$

Substituting these values in (9-6) yields the ultimate bearing capacity

$$q_u = (500)(37) + \tfrac{1}{2}(110)(6)(22) + (110)(5)(22) = 37960 \text{lb/ft}^2 \quad \text{or} \quad 18.980 \ t/\text{ft}^2$$

Comparison of the calculated values of q_u in Problems 9.14 and 9.15, for $\phi = 25°$ and $\phi = 30°$, illustrates that the ultimate bearing capacity is very sensitive to the variation of the angle of internal friction.

9.16 Determine the ultimate bearing capacity of the square footing illustrated in Fig. P-9.16.

Since the soil has a reasonably high value of cohesion and internal friction, failure of the foundation in the field would probably occur by general shear (Case 1, Table 9.2). Thus

Solution

$$q_n = 1.3cN_c + 0.4\gamma bN_\gamma + \gamma dN_q$$

From Fig. 9-3b, the dimensionless parameters for $\phi = 20°$ are

$$N_c = 17.7, \quad N_q = 7.4, \quad N_\gamma = 5$$

and

$\gamma_m = 112$ lb/cu. ft
$\phi = 20°$
$c = 600$ lb/ft^2

$d = 4$ ft

$-b = 6$ ft$-$

Figure P-9.16

$$q_u = 1.3(600)(17.7) + 0.4(112)(6)(5) + (112)(4)(7.4)$$
$$= 14800 + 1344 + 3330 = 19474 \ \text{lb/ft}^2 \quad \text{or} \quad 9.737 \ t/\text{ft}^2$$

9.17   Solve Problem 9.16, assuming a circular footing of 6-ft diameter.

Solution   To solve this problem, Terzaghi's modified equation for the ultimate bearing capacity of circular footing is used (Case 1, Table 9-2).

$$q_u = 1.3cN_c + 0.3\gamma_m bN_y + \gamma_m dN_q \tag{a}$$

The dimensionless parameters for $\phi = 20°$ are the same as in Problem 9.16. Also, the values of c, γ_m, b, and d in equation (a) above are the same. Thus

$$q_u = 1.3(600)(17.7) + 0.3(112)(6)(5) + (112)(4)(7.4)$$
$$= 19138 \ \text{lb/ft}^2 \quad \text{or} \quad 9.569 \ t/\text{ft}^2 \tag{b}$$

Slope Stability-Culmann Method

9.18   Derive Culmann's equation (9-12) for the stability analysis of slopes with a plane rupture surface.

Answer   In order to derive this relationship, reference is made to Fig. 9-5. The stability analysis assumes a shear plane $\overline{OA}$, through the toe of the slope, and requires the finding of the so-called critical angle of the slope that corresponds to the most dangerous rupture surface (maximum stress surface).
The weight of the ruptured soil wedge, $\Delta \overline{OBA}$, is

$$W = \frac{1}{2} \gamma_m(\overline{OA})h = \frac{1}{2} \gamma_m H^2 \left[\frac{\sin \beta \sin (\beta - \theta)}{\sin^2 \beta \sin \theta} \right] \tag{a}$$

The equilibrium force on plane $(\overline{OA})$ is

$$c(\overline{OA}) + N \tan \phi - T = 0 \tag{b}$$

where $c(\overline{OA})$ is the acting cohesive force, $N \tan \phi$ is the acting frictional force, and

$$(\overline{OA}) = \frac{H}{\sin \theta} \tag{c}$$

$$N = W \cos \theta = \frac{1}{2} \gamma_m H^2 \left[\frac{\sin \beta \sin (\beta - \theta)}{\sin^2 \beta \sin \theta} \right] \cos \theta \tag{d}$$

$$T = W \sin \theta = \frac{1}{2} \gamma_m H^2 \left[\frac{\sin \beta \sin (\beta - \theta)}{\sin^2 \beta \sin \theta} \right] \sin \theta \tag{e}$$

Then (b) in terms of (c), (d), and (e) is

$$c = \frac{1}{2} \gamma_m H \left[\frac{\sin (\beta - \theta)(\sin \theta - \cos \theta \tan \phi)}{\sin \beta} \right] \tag{f}$$

For the most unstable position of the rupture surface, the acting cohesion c is assumed to attain a maximum value; that is,

$$\sin(\beta - \theta)(\sin\theta - \cos\theta\tan\phi) = \text{maximum}$$

$$\frac{d}{d\theta}[\sin(\beta - \theta)(\sin\theta - \cos\theta\tan\phi)] = 0$$

Hence

$$\tan(\beta - \theta) = \frac{\sin\theta - \cos\theta\tan\phi}{\cos\theta + \sin\theta\tan\phi} = \tan(\theta - \phi) \tag{g}$$

Thus the critical angle θ_{CR} identifying the position of the most dangerous rupture surface $(\overline{OA})$ is

$$\theta_{CR} = \frac{\beta + \phi}{2} \tag{h}$$

The substitution of (h) into (f) yields the expression for the maximum cohesion, and c_{max} on the most dangerous plane $(\overline{OA})$ upon rupture is

$$c_{max} = \frac{1}{2}\gamma_m H\left[\frac{\sin^2[(\beta - \phi)/2]}{\sin\beta\cos\phi}\right] \tag{i}$$

In turn, the critical height for a known c is

$$H_{CR} = \left(\frac{2c}{\gamma_m}\right)\left[\frac{\sin\beta\cos\phi}{\sin^2[(\beta - \phi)/2]}\right] \tag{j}$$

Here equations (i) and (j) are the same as those in (9-12).

9.19   An embankment of cohesive soil is to be constructed on a 2 (vertical) to 1 (horizontal) slope. The soil has the following characteristics: cohesion $c = 800\,\text{lb/ft}^2$, angle of internal friction $\phi = 6°$, and the bulk density $\gamma_m = 110\,\text{lb/ft}^3$. Using Culmann's method (Section 9-4) and a safety factor of 2.0, determine the permissible height of the slope.

By (9-12)   *Solution*

$$H_{CR} = \frac{2c\sin\beta\cos\phi}{\gamma_m\sin^2[(\beta - \phi)/2]} = \frac{(2)(800)(0.8944)(0.994)}{(110)(0.23)} = 56.5\,\text{ft}$$

Applying the given factor of safety of 2, the allowable height is

$$H_{allow} = \frac{H_{CR}}{F_s} = \frac{56}{2} = 28\,\text{ft}$$

9.20   How deep a vertical cut can be made in the soil described in Problem 9.17 before bracing would be required? Use a factor of safety of 2.0.

Solution From (9-12), and after it is modified for $\beta = \pi/2$ (vertical cut), the critical height is

$$H_{CR} = 4\left(\frac{c}{\gamma_m}\right) \tan\left(\frac{\pi}{4} + \frac{\phi}{2}\right) = 8\left(\frac{800}{110}\right) \tan\left(45 + \frac{6}{2}\right) = 29.1 \tan 48° = 32.5 \text{ ft}$$

Thus a bracing will be required if the height of cut exceeds

$$H_{\text{allow}} = \frac{H_{CR}}{F_s} = \frac{32}{2.0} = 16 \text{ ft}$$

Petterson Method (Circular Arc or Swedish Method)

9.21 Describe how you determine the position of the center of the critical circle around which the unstable sector (Fig. 9-6) would rotate.

Answer The center of the most critical circle can be found by trial and error as follows:

1. A grid pattern, such as that shown in Fig. P-9.21, is established alongside the slope.

2. Each of the corners of the grid is then assumed to be a center for a trial slip circle; and by using equation (9-14), the factor of safety F_s for each circle is determined and recorded on the grid pattern as shown in Fig. P-9.21.

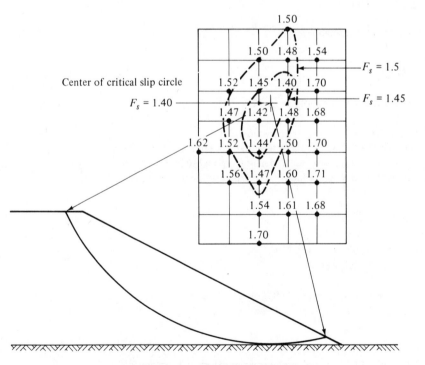

Figure P-9.21   (T. W. Lambe and Robert Whitmann, *Soil Mechanics*, Wiley, New York, 1969, p. 369.)

3. After several points have been established, points of equal F_s values are joined, which is similar to drawing contours of F_s values. These contours are approximately elliptical in shape.

4. The center of these elliptical contours indicates where the center of the critical slip circle would be.

In order to find a reasonable position of the center of the first trial slip circle, Table P-9.21 can be used. The table was compiled by Taylor* from a great many investigations, using an analytical solution of the ϕ circle method on the basis of total stress (see Problem 9.34).

9.22   An embankment built from a purely cohesive soil has a cross section as shown in Fig. P-9.22. The material has a density of 120 lb/ft³ and a cohesion c of 1000 lb/ft². For the trial slip circle shown, determine the factor of safety.

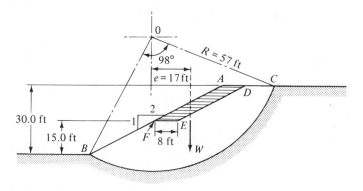

Figure P-9.22

First, the area of the sliding sector $ABCA$ is measured by a planimeter and in this *Solution* case is:

$A = 1550$ ft²

The weight of a 1-ft section of the sliding mass is

$W = A\gamma_m = (1550)(120) = 186,000$ lb

To find the distance e (Fig. P-9.22) of the line of action of the weight W from the center of rotation, a piece of cardboard may be cut to the shape of the sector and suspended in turn from two points. Vertical lines drawn through the points of suspension will intersect at the center of gravity. In this case, the line of action of W is at a distance $e = 17$ ft.

Then, in order to determine the factor of safety against sliding, the following quantities must be computed first.

*D. W. Taylor, *Fundamentals of Soil Mechanics*, Wiley, New York, 1948, p. 460.

TABLE P-9.21

Slope β	Angle of Friction ϕ	Angle for Setting Out Center of Critical Circle		Depth Factor D	Stability Number $\dfrac{c}{F\gamma H}$
		ψ	θ		
90	0	47.6	30.2	–	0.261
	5	50.0	28.0	–	0.239
	10	53.0	27.0	–	0.218
	15	56.0	26.0	–	0.199
	20	58.6	24.0	–	0.182
	25	60.0	22.0	–	0.166
75	0	41.8	51.8	–	0.219
	5	45.0	50.0	–	0.195
	10	47.5	47.0	–	0.173
	15	50.0	46.0	–	0.152
	20	53.0	44.0	–	0.134
	25	56.0	44.0	–	0.117
60	0	35.3	70.8	–	0.191
	5	38.5	69.0	–	0.162
	10	41.0	66.0	–	0.138
	15	44.0	63.0	–	0.116
	20	46.5	60.4	–	0.097
	25	50.0	60.0	–	0.079
45	0	(28.2)	(89.4)	(1.062)	(0.170)
	5	31.2	84.2	1.026	0.136
	10	34.0	79.4	1.006	0.108
	15	36.1	74.4	1.001	0.083
	20	38.0	69.0	–	0.062
	25	40.0	62.0	–	0.044
30	0	(20.0)	(106.8)	(1.301)	(0.156)
	5	(23.0)	(96.0)	(1.161)	(0.110)
	5	20.0	106.0	1.332	0.110
	10	25.0	88.0	1.092	0.075
	15	27.0	78.0	1.038	0.046
	20	28.0	62.0	1.003	0.025
	25	29.0	50.0	–	0.009
15	0	(10.6)	(121.4)	(2.117)	(0.145)
	5	(12.5)	(94.0)	(1.549)	(0.068)
	5	11.0	95.0	1.697	0.070
	10	(14.0)	(68.0)	(1.222)	(0.023)
	10	14.0	68.0	1.222	0.023

For ψ, θ refer to Fig. P-9.6. For D refer to Table P-9.29.

NOTE: Figures in parentheses are for most dangerous circle through the toe when a more dangerous circle exists that passes below the toe. From P. L. Capper, W. F. Cassie, and J. D. Geddes, *Problems in Engineering Soils*, E. & F. N. Spon, London, 1966, p. 127.

a) Disturbing moment:

$$M_d = We = (186,000)(17) = 3,162,000 \text{ lb-ft}$$

(b) Restraining moment:

$$M_r = cL = cr^2\theta = (1000)(57)^2\left[\left(\frac{98}{180}\right)(\pi)\right] = 555,434 \text{ lb-ft}$$

Hence the factor of safety is

$$F_s = \frac{\text{restraining moment}}{\text{disturbing moment}} = \frac{M_r}{M_d} = \frac{555,434}{3,162,000} = 1.76$$

9.23 In Problem 9.22, what would be the factor of safety if the shaded portion of the embankment shown in Fig. P-9.22 is removed?

Area A' of portion $ADEF$ (Fig. P-9.22) is   *Solution*

$$A' = (3)(8) = 24 \text{ ft}^2$$

Assuming a 1-ft section of the embankment, the weight of the removed portion is

$$W' = (A')(\gamma_m) = (24)(120) = 2880 \text{ lb}$$

and the eccentricity of this removed section from the center of rotation is

$$e' = 23 \text{ ft}$$

Thus the relief moment is

$$(W')(e') = (2880)(23) = 66,240 \text{ lb-ft}$$

and the factor of safety is

$$F_s = \frac{M_r}{M_d} = \frac{5,554,346}{3,162,000 - 66,240}$$

$$= 1.80$$

The comparison of Problems 9.22 and 9.23 shows that the removal of the shaded area, from the embankment, will result in an increased factor of safety against sliding along the assumed circle.

9.24 When treating slope stability problems in purely cohesive soils, a tension crack usually develops at some distance from the top of the slope, as shown in Fig. P-9.24. How would you determine the depth of such a crack, and how does it influence the analysis of the stability of the slope?

The depth of the crack in a purely cohesive soil h_c (Fig. P-9.24) is   *Solution*

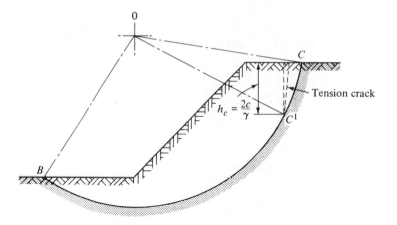

Figure P-9.24

$$h_c = \frac{2c}{\gamma} \tag{a}$$

The influence of such a tension crack evolves from the fact that the slip circle will pass through it, but no cohesive forces can develop to the depth of this crack. Thus the slip surface resisting the rotation will be BC' instead of BC (Fig. P-9.24), which, in turn, results in a reduced factor of safety. It is also highly possible that such a crack may be filled with water, which would exert a hydrostatic pressure on the walls of the crack. This situation will add to the disturbing moment around the center of rotation O, thereby resulting in a further reduction of the factor of safety.

9.25 An embankment built from a purely cohesive soil has the cross section shown in Fig. P-9.25. The embankment material has a density $\gamma_m = 120 \, \text{lb/ft}^3$ and a

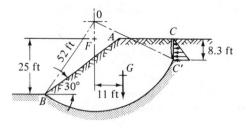

Figure P-9.25

cohesion $c = 500 \, \text{lb/ft}^2$. For the trial slip circle shown, the area was measured by a planimeter and was found to be 1700 ft²; the center of gravity was found to be at point G, as shown in Fig. P-9.25. Determine the factor of safety for this slip circle, assuming that the embankment serves as a reservoir bank with water level at the top of the bank. (Neglect the effect of any tension cracks.)

When the water level is at the top of the bank, the embankment material is sub- *Solution* merged. Thus to determine the factor of safety F_s for this slip circle, the following parameters must be determined first (considering 1 ft depth of the embankment).

Disturbing moment:

$$M_d = (1700)(120 - 62.4)(11)$$
$$= 1,077,120 \text{ lb-ft}$$

Resisting moment:

$$M_r = (AC')(c)(R) = \left[(52)(95)\left(\frac{\pi}{180}\right)\right](500)(52) = 2,240,420 \text{ lb-ft}$$

Hence the factor of safety is

$$F_s = \frac{M_r}{M_d} = \frac{2,240,420}{1,077,120} = 2.08$$

9.26 What will be the factor of safety in Problem 9.25 if the reservoir is emptied? Assume that a tension crack $C - C'$ (Fig. P-9.25) will develop and will remain filled with water after the reservoir has been drained.

The depth of the tension crack $C - C'$ (Fig. P-9.25) is computed by (a) of Problem *Solution* 9.24 as

$$h_c = \frac{2c}{\gamma} = \frac{(2)(500)}{(120)} = 8.33 \text{ ft}$$

and from Fig. P-9.25,

$$\overline{OF} = 15.8 \text{ ft}$$

Thus the disturbing moment

M_d = moment of the weight of rotating slice + moment of the water pressure in crack, both about O

$$= [(1700)(120)(11)] + [(\tfrac{1}{2})(62.4)(8.3)^2(15.8 + 5.5)] = 2,244,000 + 45781$$
$$= 2,289,781 \text{ lb-ft}$$

The resisting moment M_r will remain the same as that calculated in Problem 9.25; that is,

$$M_r = 2,240,420 \text{ lb-ft}$$

The factor of safety is then

$$F_s = \frac{2,240,420}{2,289,781} = 0.978$$

Since the factor of safety is less than 1.0, after the reservoir was drained, the embankment becomes unsafe against failure by sliding.

Fellenius Method

9.27 Show how the stability analysis along a slip plane in a frictional (c-ϕ) soil is performed by the Fellenius method.

Answer The shear strength of a mass of a frictional soil will vary with the effective normal pressure on the surface of sliding. For such types of soils, the stability analysis is carried out by the method of slices, in which the soil profile inside the assumed slip circle is divided into a convenient number of strips of equal width (Fig. P-9.27).

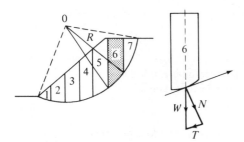

Figure P-9.27

Assuming that each strip acts independently as a column of soil with a unit thickness, the forces between the slices may be neglected.

Consider one strip—for example, strip No. 6. The weight W of the slice is assumed to pass through the midwidth point of this slice. This weight W is then resolved into two components:

1. N, normal to the arc of the slip.
2. T, tangential to the arc of the slip.

Depending on the position of the slice with respect to the center of rotation, the tangential component causes a disturbing or resisting moment, and thus is designated as positive or negative. The algebraic summation of T will always be positive and can be used to calculate the net disturbing moment as follows:

For one strip, the disturbing moment about the center O is

$$M_d' = \pm(T)(R)$$

and the total disturbing moment for the entire slip surface is

$$M_d = R[\sum (+T)] \tag{a}$$

Also, for one strip with an arc length of s, the resisting moment is

$$M_r' = (cs + N \tan \phi)R$$

Thus the total resisting moment is

$$M_r = R(cR\theta + \tan\phi \sum N) \qquad (b)$$

Hence the factor of safety can be expressed as

$$F_s = \frac{M_r}{M_d} = \frac{c(R\theta + \tan\phi \sum N)}{\sum(+T)} \qquad (c)$$

9.28 An embankment built of a homegeneous silty-clay soil has a cross section as shown in Fig. P-9.28. The soil has the following characteristics: cohesion $c = 450$ lb/ft², angle of internal friction $\phi = 8°$, and density $\gamma_m = 110$ lb/ft³. Assuming that a tension crack will develop at the top of the bank, what is the factor of safety against failure by sliding along the slip circle shown in Fig. P-9.28?

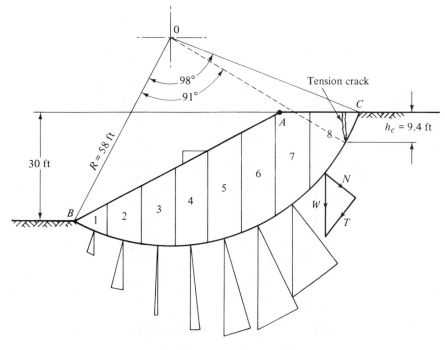

Figure P-9.28

For a frictional soil, the depth of the tension crack is expressed as *Solution*

$$h_c = \frac{2c\sqrt{N_\phi}}{\gamma_m} = \frac{2[450\tan(45° + \phi/2)]}{110} = \frac{2[450(\tan 49°)]}{110} = 9.4 \text{ ft}$$

In Fig. P-9.28, the slip circle is divided into eight strips, each about 10 ft wide. The 10-ft width is arbitrary and chosen for convenience only. Next, assuming a 1-ft thickness of the sliding sector, the weight of each slice is computed by multiplying the area of each slice by the unit weight.

The weight of each slice is then plotted to a convenient scale vertically underneath at the location of the slice centroid. Next the normal and tangential forces are drawn to complete the force diagram for each strip, as shown in Fig. P-9.28. Using these relationships, Table P-9.28 can be drawn up.

To determine the factor of safety F_s, the following quantities are calculated:

$$cR\theta = (450)(58)\left[\left(\frac{91}{180}\right)(\pi)\right] = 41,432$$

$$\Sigma N \tan \phi = (152,300)(\tan 8°) = 21,398$$

By equation (c) of Problem 9.27, the factor of safety is

$$F_s = \frac{c(R\theta + \tan \phi \, \Sigma N)}{\Sigma(+T)} = \frac{(41,432) + (21,398)}{(47,200)} = 1.33$$

TABLE P-9.28

Slice Number	Weight W (lb)	Components	
		Tangential T (lb)	Normal N (lb)
1	5,150	−1,900	4,700
2	14,380	−3,300	14,000
3	21,300	−1,200	21,200
4	26,800	+3,000	26,600
5	30,800	+8,600	28,600
6	29,900	+14,100	27,500
7	26,100	+16,400	20,400
8	14,600	+11,500	9,300
		$\Sigma(+T) = +47,200$ lb	$\Sigma N = 152,300$ lb

Taylor Method

9.29 Discuss the use of Taylor's method in the analysis of slope stability problems.

Answer In analyzing the stability of a finite slope, the weight (or density) of the soil in the sliding sector generally contributes to the disturbing forces. On the other hand, the cohesion c contributes to the resisting forces. Hence the critical height H_{CR}, which is the maximum possible height of a slope, is directly proportional to the unit cohesion c and inversely proportional to the density γ_m. This critical height also depends on the angle of friction ϕ and the angle of slope β.

Thus a mathematical expression of H_{CR} may be stated as

$$H_{CR} = \frac{c}{\gamma_m} f(\phi, \beta) \tag{a}$$

where $f(\phi, \beta)$ is a dimensionless function. Terzaghi (1954) called this function "the stability factor." Taylor (1937) expressed it as a reciprocal of a certain dimensionless number, called the stability number S_n.

Writing the above expression (a) according to Taylor's method yields

$$H_{CR} = \frac{c}{\gamma_m S_n} \quad \text{or} \quad S_n = \frac{c}{\gamma_m H_{CR}} \tag{b}$$

But since the factor of safety F_c is usually used with respect to cohesion, equation (b) is written as

$$S_n = \frac{c}{F_c \gamma_m H} = \frac{c_m}{\gamma_m H} \tag{c}$$

where H = height of a stable slope assuming a factor of safety F_c $(H < H_{CR})$

c_m = mobilized unit cohesion necessary for equilibrium of the slope of height H.

Taylor compiled the data shown in Tables P-9.21 and P-9.29, where S_n is shown as a function of ϕ and β. For the the application of this method, see Problem 9.30.

It should be noticed, however, that Taylor's stability number gives a factor of safety F_c with respect to cohesion only. If a true factor of safety F_s to both cohesion and friction is required, $\tan \phi$ must also be divided by F_s, as shown in Problem 9.31.

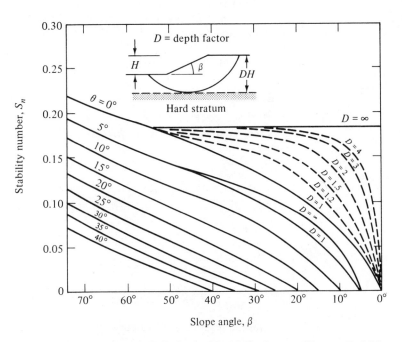

(P. L. Capper and W. F. Cassie, *The Mechanics of Engineering Soils*, E. & F. N. Spon, London, 1969, p. 136.)

Figure P-9.29

9.30 A highway cut is to be constructed in a silty-clay soil to a depth of 25 ft and with a 1 : 1 slope. The soil has a cohesion $c = 700$ lb/ft², angle of friction $\phi = 10°$, and natural density $\gamma_m = 110$ lb/ft³. Determine the factor of safety F_c with respect to cohesion, plus the critical height H_{CR} of the slope of this soil.

Solution For $\beta = 45°$ and $\phi = 10°$, the stability number taken from Table P-9.21 or Table P-9.29 is

$$S_n = 0.108$$

Then, using (c) of Problem 9.29,

$$S_n = \frac{c}{F_c \gamma_m H} \quad \text{or} \quad 0.108 = \frac{700}{F_c(110)(25)}$$

from which

$$F_c = 2.35$$

By (b) of Problem 9.29, the critical height of the slope H_{CR} is

$$H_{CR} = \frac{c}{\gamma_m S_n} = \frac{700}{(110)(0.108)} \cong 59 \text{ ft.}$$

9.31 Determine the safe depth of a cut, at a slope angle $\beta = 40°$, which is to be constructed in a silty-clay soil with cohesion $c = 700$ lb/ft², angle of friction $\phi = 15°$, and natural density $\gamma_m = 112$ lb/ft³. The required true factor of safety is 1.5.

Solution The mobilized angle of friction by definition is

$$\phi_m = \frac{\phi}{F_s} = \frac{15}{1.5} = 10°$$

Then, from Table P-9.21, for $\beta = 40°$ and $\phi_m = 10°$, the stability number is

$$S_n = 0.098$$

By (c) of Problem 9.29,

$$S_n = \frac{c}{F_s \gamma_m H}$$

from which

$$H = \frac{c}{F_s \gamma_m S_n} = \frac{700}{(1.5)(112)(0.098)} = 42.5 \text{ ft}$$

Thus the required depth of cut in this material, using a factor of 1.5, is about 42 ft.

9.32 A dam 20 ft deep was cut at a slope of 1 : 1 through a soil that has a cohesion $c = 500$ lb/ft², angle of internal friction $\phi = 10°$, and saturated density $\gamma_{sat} = 122$

lb/ft³. Determine the factor of safety with respect to cohesion when the reservoir is full to the top level of the cut. Also, determine the factor of safety in case of a sudden drawdown and compare with the previously determined value when the reservoir is full.

The submerged unit weight of the soil is   *Solution*

$$\gamma_{\text{sub}} = \gamma_{\text{sat}} - \gamma_0 = 122 - 62.4 = 59.6 \text{ lb/ft}^3$$

When the reservoir is full, and for $\beta = 45°$ and $\phi = 10°$, the stability number from Table P-9.21 is

$$S_n = 0.108$$

By equation (c) of Problem 9.29,

$$F_c = \frac{c}{\gamma_{\text{sub}} H S_n} = \frac{500}{(59.6)(20)(0.108)} = 3.89$$

When a rapid drawdown occurs, the friction ϕ will be reduced to the mobilized friction ϕ_m as

$$\phi_m = \frac{\gamma_{\text{sub}}}{\gamma_{\text{sat}}} \phi = \frac{59.6}{122}(10) = 4.88°$$

For $\beta = 45°$ and $\phi_m = 4.88°$, the stability number from Table P-9.21 is

$$S_n = 0.137$$

By equation (c) of Problem 9.29,

$$F_c = \frac{c}{(\gamma_{\text{sat}})(H)(S_n)} = \frac{500}{(122)(20)(0.137)} = 1.49$$

It is clear from this example that the rapid drawdown of water from the reservoir resulted in a reduction of the factor of safety for the bank by more than 50 percent.

9.33 A highway cut is made in a soil that has a cohesion $c = 700 \text{ lb/ft}^2$, angle of internal friction $\phi = 12°$, and a natural density $\gamma_m = 112 \text{ lb/ft}^3$. The slope of the cut is 30° and it has a depth of 50 ft. Determine the factor of safety of the slope.

If the friction is fully mobilized, the value of the stability factor S_n (by interpolation) *Solution* from Table P-9.29 for $\phi = 12°$ and $\beta = 30°$ is

$$S_n = 0.063$$

By equation (c) of Problem 9.29, the factor of safety for this value of S_n is

$$F_c = \frac{c}{S_n \gamma_m H} = \frac{700}{(0.063)(112)(50)} = 1.98 \qquad\qquad (a)$$

But since the friction is not fully mobilized, the actual factor of safety must be less than that calculated in (a). The actual value of F_s can be found by successive approximations, using the value of F_c calculated in (a) as a guideline to a starting value and assuming a value of $F_s = 1.6$. Then the mobilized angle of friction ϕ_m becomes

$$\tan \phi_m = \frac{\tan 12°}{1.6} = \frac{12}{1.6} = 7.5°$$

Then from the table, for $\phi = 7.5°$, the stability number is

$$S_n = 0.092$$

Hence the second approximation of the factor of safety is

$$F_s = \frac{c}{S_n \gamma H} = \frac{700}{(0.092)(112)(50)} = 1.36$$

Since this value does not agree with the assumed value of $F_s = 1.6$, a new value should be attempted (e.g., 1.4). Following the same procedures as above, calculations yield a factor of safety of 1.47, which is close to the assumed value of 1.40. Thus the final value of the factor of safety F_s is $\cong 1.45$.

Friction Circle Method

9.34   Explain how the friction circle method is used in solving slope stability problems.

Answer   The friction circle method is widely used for the solution of slope stability problems in c-ϕ soils, or soils possessing both cohesion and internal friction. The use of this method is summarized in the following steps:

1.   Assuming that the slip surface is along the arc BC (Fig. P-9.34a), lay off the line of action of the weight of the sliding sector W.

2.   Determine the line of action of C which is the resultant of all the cohesive forces along the surface of sliding BC. The position of C is determined as follows:

(a)   Let l be the length of the arc BC and l' be the length of the chord of this arc. The moment about the center of rotation O of the mobilized cohesion c_m (only part of the ultimate cohesion c) acting on an element of length δl is

$$[(c_m)(\delta l)(R)]$$

Assuming that cohesion is uniformly distributed over the sliding surface BC, the resultant of the cohesive forces on all such elements is a force $C = (c_m)(l')$, acting at a perpendicular distance a from O, where

$$c_m l' a = c_m l R$$

from which

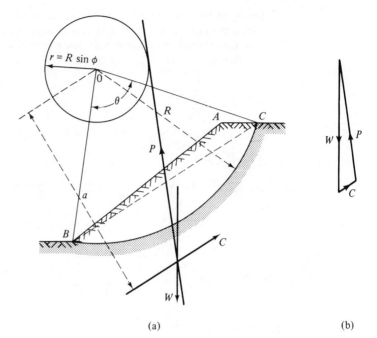

(a)   (b)

Figure P-9.34

$$a = \frac{lR}{l'} = \frac{R\theta}{2} \sin \frac{\theta}{2} \tag{a}$$

3. Determine the intersection of the line of action of W and C.
4. Draw a circle with a radius $r = R \sin \phi$ with center at O.
5. Draw a line through the point of intersection of the lines of action of W and C [determined in (3)] which is a tangent to the friction circle drawn in (4). This step will determine the direction of the resultant force P (Fig. P-9.34a).
6. Construct the force triangle as shown in Fig. P-9.34b. Since W is known in direction and magnitude and C and P are known in direction, the force system may now be solved and the magnitude of C found.
7. From (6) it will then be easy to determine the value of the mobilized cohesion c_m necessary for equilibrium.
8. The ratio of the ultimate cohesion c to the mobilized cohesion c_m may thus be determined, giving the value of the factor of safety F_c.

In the preceding solution it was assumed that all available frictional resistance ϕ is fully mobilized. However, such may not be true and only a portion of the ultimate frictional resistance may be fully mobilized. To consider this situation, the radius of the friction circle is redrawn at a reduced radius r' of $R \sin \phi_m$, where ϕ_m, is such that $F_c \tan \phi_m = \tan \phi$. The computations and graphical solution as explained previously in steps (1) to (8) must then be repeated until the two values of factor of safety for ϕ and c are nearly equal (see Problem 9.35).

9.35 A highway embankment has a cross section as shown in Fig. P-9.35a. Determine the factor of safety with respect to cohesion, and also the true factor of safety, for the assumed slip circle shown. The soil has the following properties: cohesion c = 500 lb/ft², angle of internal friction $\phi = 12°$, and density $\gamma_m = 112$ lb/ft³.

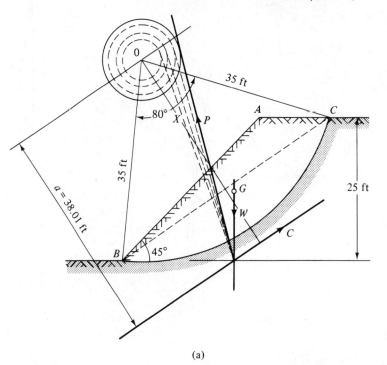

(a)

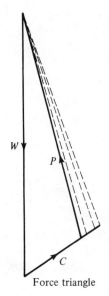

Force triangle

(b)

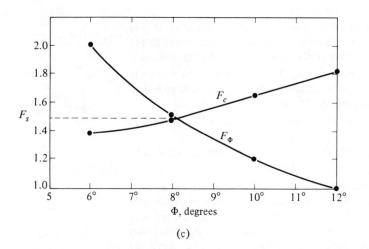

(c)

Figure P-9.35

Following the steps enumerated in Problem 9.34, and referring to Fig. P-9.35, the *Solution*
quantities below are determined first:

Angle $BOG = 80° = 1.3963$ radians

Length of arc BC, $l = R\theta = (35)(1.3963) = 48.87$ ft

Length of chord BC, $l' = 45.0$ ft

By equation (a) of Problem 9.34 (step 2), the distance a is

$$a = \frac{lR}{l'} = \frac{(35)(48.87)}{45.0} = 38.01 \text{ ft}$$

Then the force C is drawn (as shown in Fig. P-9.35a) parallel to the chord and at
a distance $a = 38.01$ ft away from the arc center O.
For a section of the embankment 1 ft deep, the weight W of the sliding mass is

$$W = (\text{area } BCA)(\gamma_m)(1.0) = (408.3)(112)(1.0) = 45730 \text{ lb}$$

The force W is then drawn vertically through the centroid of the mass as shown in
Fig. P-9.35a, and its intersection with C is located. The friction circle is next drawn
with a radius

$$r = R \sin \phi = (35)(\sin 12°) \cong 7.3 \text{ ft}$$

The force P is then drawn (Fig. P-9.35a) through the point of intersection of W and
C, tangent to the friction circle. The force diagram of P, W, and C is next plotted
as shown in Fig. P-9.35b. From this force triangle, $C = 12,370$ lb and

$$c_m = \frac{Ca}{Rl} = \frac{(12370)(38.01)}{(35)(48.87)} = 274.9 \text{ lb/ft}^2$$

Finally,

$$F_c = \frac{500}{274.9} = 1.82$$

TABLE P-9.35

Mobilized Friction Angle ϕ_1 (degree)	$R \sin \phi_1$	$F_\phi = \dfrac{\tan 12°}{\tan \phi_1}$	C (lbs)	$c_1 = \dfrac{C}{45.0}$ (lb/ft²)	$F_c = \dfrac{c}{c_1}$
12	7.28	1.000	12370	274.9	1.82
10	6.08	1.206	13390	297.5	1.68
8	4.87	1.512	15120	335.9	1.49
6	3.67	2.022	16140	358.8	1.39

To find the true factor of safety (assuming it is the same for friction as for cohesion), repeat the preceding construction for friction angles ϕ of 6, 8, and 10 degrees, as shown in Table P-9.35 and plot F_ϕ and F_c against, ϕ_1, as shown in Fig. P-9.35c. The curves intersect at $F_\phi = F_c = 1.50$ and give the required factor of safety $F_s = 1.5$.

SUPPLEMENTARY PROBLEMS

Modes of Failure—Force Interaction

9.36 If a specimen of silty-clay soil is subjected to an undrained triaxial test with a confining pressure p_3 of 40 lb/in.², cohesion $c = 10$ lb/in.², and angle of internal friction $\phi = 12°$, compute the total axial pressure p_1 at which failure would occur.

Answer Total axial pressure, $p_1 = 143.5$ lb/in².

9.37 A quick undrained triaxial test was performed on a saturated clay. The confining pressure p_3 is 12 lb/in.² and the total vertical pressure at which the sample failed was 35 lb/in.². Compute the soil parameters c and ϕ.

Answer Cohesion, $c = 11.5$ lb/in.², and angle of internal friction, $\phi = 0°$.

Plate-Loading Test

9.38 A plate-loading test conducted by a 30-in. diameter plate yielded the results shown in Fig. P-9.38. Determine the modulus of subgrade reaction, k.

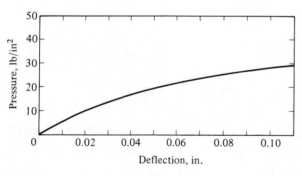

Figure P-9.38

Answer Using the P.C.A. method:
The modulus of soil reaction, $k = 400$ lb/in.³
The modulus of soil reaction (corrected for plate bending, Fig. P-9.6c),

$$k_b = 325 \text{ lb/in.}^3$$

Using the Corps of Engineers method:

$k = 500 \text{ lb/in.}^3$
$k_b = 380 \text{ lb/in.}^3$

9.39 Develop an expression for the modulus of soil reaction k (as determined by the plate-loading test) in terms of the modulus of elasticity E of the soil.

$E = 1.18 \, (pa/d)$, where p is the pressure at a settlement d of the soil under a plate *Answer*
with a radius a.

9.40 A plate-loading test using a 30-in. diameter plate was found to have an average plate settlement of 0.2 in. under a pressure of 40 lb/in.² Compute approximate value of the modulus of elasticity of this subgrade.

$E = 3540 \text{ lb/in.}^2$ *Answer*

9.41 The results shown in Fig. P-9.41a were obtained in a field plate-loading test. Figure P-9.41b represents the deformation-time curves for consolidation tests under a uniform pressure of 10 lb/in.², performed on typical soil samples from the field, at natural moisture content and at saturated state. Using the data presented in these figures, determine the soil's modulus of subgrade reaction k according to the Corps of Engineers procedures. Also, make all necessary corrections for the calculated value.

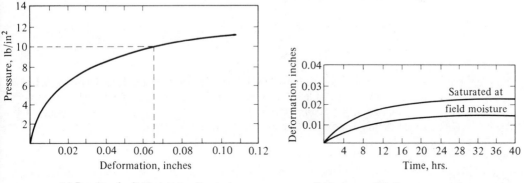

(a) Results of a field plate-loading test

(b) Lab consolidation curves under a pressure of
10 psi for the same soil subgrade

Figure P-9.41

The uncorrected value of subgrade reaction modulus, $k = 154 \text{ lb/in.}^3$ *Answer*
    The value of subgrade reaction modulus, corrected for plate bending, $k_b = 150$
lb/in.³
    The value of subgrade reaction modulus further corrected for saturated condition
of soil, $k_s = 81.5 \text{ lb/in.}^3$

Prandtl Method

9.42 Determine the bearing capacity of a foundation supporting soil that has the following properties: angle of internal friction $\phi = 20$ and cohesion $c = 160$ lb/ft². Assume soil to be weightless to the depth of the footing.

Answer Ultimate bearing capacity (using Prandtl equation)

$$q_u = 1500 \text{ lb/ft}^2 \text{ or} \cong 0.750 \, t/\text{ft}^2$$

Terzaghi Method

9.43 Consider a strip footing 5-ft wide and constructed with the lower surface of footing at a depth of 4 ft from the ground surface. The foundation soil has a bulk density $\gamma_m = 112$ lb/ft³, cohesion $c = 600$ lb/ft², and angle of internal friction $\phi = 21°$. Determine the ultimate bearing capacity q_u.

Answer The ultimate bearing capacity

$$q_u = 21160 \text{ lb/ft}^2 \text{ or} \cong 10.58 \, t/\text{ft}^2$$

9.44 A shallow strip footing is constructed in a purely cohesive soil (clay) at a depth of 5 ft. It must carry a pressure of 5 t/ft^2 safely. If the foundation soil has a bulk density γ_m of 110 lb/ft² and cohesion c of 1200 lb/ft², determine the minimum width of the strip footing, assuming a factor of safety of 2.0.

Answer Allowable bearing capacity $q_{\text{allow}} = 1.847 \, t/\text{ft}^2$

Width of the strip footing,

$$b = 2.71 \text{ ft or} \cong 3.0 \text{ ft}$$

9.45 A rectangular footing 25 by 8 ft is to be constructed in a silty-clay soil at a depth of 5 ft. The soil has the following properties: angle of internal friction $\phi = 15°$, cohesion $c = 1500$ lb/ft², and bulk density $\gamma_m = 124$ lb/ft³. Assuming a safety factor of 2.0, determine the allowable bearing capacity.

Answer Allowable bearing capacity,

$$q_{\text{allow}} = 6.6 \, t/\text{ft}^2$$

Slope Stability Analysis

9.46 Using Culmann's method, determine the critical height of a vertical cut in a purely cohesive soil that has a cohesion c of 1500 lb/ft² and a bulk density γ_m of 112 lb/ft³.

Answer Critical height H_{CR} (assuming a factor of safety of 1.0) $\cong$ 53 ft.

9.47 An embankment, as shown in Fig. P-9.47, is resting on a stiff layer at the surface of the ground. The characteristics of the embankment material are as follows:

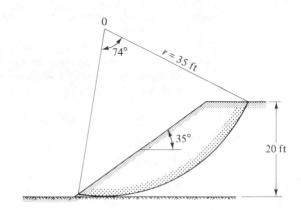

Figure P-9.47

density $\gamma_m = 120$ lb/ft^3, cohesion $c = 600$ lb/ft^2, and angle of internal friction $\phi = 15°$. Determine the factor of safety for the slip circle shown in Fig. P-9.47. (Neglect the effect of tension cracks.)

$F_s = 2.14.$ *Answer*

9.48   An earth dam has a cross section as shown in Fig. P-9.48. The dam has the following characteristics: cohesion $c = 200$ lb/ft^2, angle of internal friction $\phi = 30°$.

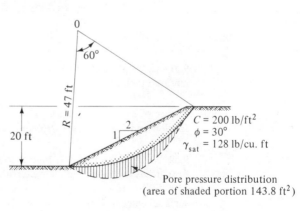

Figure P-9.48

The soil is saturated and has a bulk density γ_m of 128 lb/ft^3. The pore-water pressure on the slip surface (found from a flow net) is represented by the ordinates of the hatched area (from the slip surface to the dotted curve) on Fig. P-9.48. By dividing the slip sector into a convenient number of slices, determine the factor of safety of the slope under conditions of a steady seepage. (Neglect the effect of tension cracks.)

Answer The pore-water pressure u in this case will be acting upward normal to the slip surface and therefore has no driving moment around the center of rotation O of the slice. Thus the N component of the weight of the slice is reduced to $(N - u)$ because of the pore-water pressure. By dividing the sliding sector into a convenient number of slices and introducing the modified normal pressure $(N - u)$ into equation (c) of Problem 9.26, we have

$$F_s = \frac{cR\theta + \tan\phi \; \Sigma(N - u)}{\Sigma(+T)}$$

Applying the above procedure to this problem yields

$$F_s = 1.1$$

9.49 Solve Problem 9.48 without using the pore-water pressure diagram given in the problem; instead use the alternative approximate method in which the weight of the submerged density of the soil is used to calculate the weight of the sliding sector of the slope.

Answer Submerged density

$$\gamma_{\text{sub}} = (128 - 62.4) = 65.6 \; \text{lb/ft}^3$$

$$F_s = 0.92 \; \text{(Compare this value with that produced in Problem 9.48.)}$$

9.50 A highway cut 90 ft deep has a slope of 1 (vertical) to 2 (horizontal). The properties of the soil are: cohesion $c = 500 \; \text{lb/ft}^2$, internal friction $\phi = 21°$, and density $\gamma_m = 110 \; \text{lb/ft}^3$. Using Taylor's charts, determine the factor of safety for the slope.

Answer $F_s = 1.38$

9.51 Solve Problem 9.28, using the friction circle method. (Neglect the effect of tension cracks.)

Answer Factor of safety $F_s = 1.30$. (This is the true factor of safety, which is the same for friction and cohesion; see Problem 9.35.)

9.52 A highway cut 40 ft high is to be constructed in a purely cohesive soil ($\phi = 0$) that has the following characteristics: density $\gamma_m = 120 \; \text{lb/ft}^3$ and cohesion $c = 1100 \; \text{lb/ft}^2$. If a stiff stratum exists at a depth of 50 ft below the original ground surface, determine the safe slope of cut, allowing for a factor of safety of 1.50.

Answer Depth factor (from Taylor's table),

$$D = \frac{50}{40} = 1.25$$

Stability number,

$$S_n = \frac{c}{F_s \gamma_m H} = 0.152$$

Then, from Taylor's charts and for $\phi = 0$ and $D = 1.25$, the required slope angle $\beta \cong 30°$.

frost
action
10

10.1 INTRODUCTION

General

When the surface temperature falls below the freezing point, $0°\,C = 32\;°F$ (Table 4-2), *freezing of the soil water* may occur in the subgrade. Such an occurrence may be either permanent or temporary (cyclic), and it may reach shallow or great depths of the subgrade.

In Arctic and Antarctic regions, the soil remains frozen throughout the year to great depths (up to 1000 ft and more) in a condition called *permafrost*. A special characteristic of permafrost is the thaw in summer months, which converts the top layer of the ground into soupy quagmire, a highly unstable material

In other regions, where the freezing occurs during the winter season only, the frost penetrates a few inches or feet below the ground surface and retreats with the change in season.

The study of the permafrost action is a specialized part of engineering soil mechanics not covered in this book; the study of the shallow-depth frost penetration and its effects in the subgrade are important factors in foundation design and as such are the prime concern of the subsequent discussion.

Frost Action

The prerequisites for the freezing of soil water are the duration of the freezing temperature for a reasonable period of time (longer than 3 days), an adequate supply of moisture in the soil (free water table close to the surface), and a frost-susceptible soil.

When water changes from a liquid to a solid state, its volume expands about 10 percent; and in a saturated soil with a void ratio of 0.6 to 0.8, the change in volume is about 4 to 6 percent. If melting takes place, an equivalent decrease in volume must occur.

Frost Damage

Damage to shallow foundations, small bridges, culverts, walls, highways, and airfield pavements may occur as a result of the frost action classified as the frost heave or the spring breakup.

1. The *frost heave* occurs as a result of suction associated with the freezing process, causing water to migrate into the frozen soil zone and form ice lenses. These lenses will continue to grow in size as a result of a continuous supply of moisture from available sources in the soil system. This process will ultimately produce heavy pressures and large displacements at the contact planes of structure and the supporting soil system, leading eventually to a structural damage or failure.

2. The *spring breakup* takes place during the frost-melting period. The moisture accumulates in the previously frozen zone of soil, after thawing of the ice lenses causing a lowering of the bearing capacity of the soil system for a period of time, also leading to a structural damage or a possible failure.

10.2 *TEMPERATURE PROFILES*

General

Since the temperature in soil varies with depth as well as with time, two temperature profiles (charts) are used for the graphical representation of this variation, the *depth profile* and the *annual profile*.

Depth Profile

The variation of temperature with depth for any given time in a given soil system is represented graphically by the temperature depth profile which penetrates two layers: (a) the active layer (subject of freezing and melting) and (b) the inactive layer (independent of surface freezing).

Two typical cases of temperature depth profile are shown in Fig. 10-1.

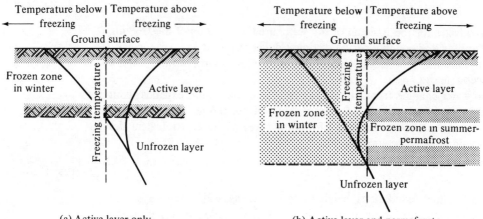

(a) Active layer only (b) Active layer and permafrost

Figure 10-1   *Depth temperature profiles*

(K. Terzaghi, "Permafrost," J. Boston Soc. Civ. Eng. Vol. 39, No. 1, 1952.)

Annual Profile

The annual histogram of the surface temperature variations with time for a given geographic location is represented graphically by the temperature annual profile. The horizontal coordinate of this profile is the number of days counted from an arbitrarily selected datum, and the vertical coordinate is the mean daily temperature.

Freezing Index

The area enclosed by the annual temperature profile and its time axis is called the temperature index, which, for the area below this axis, is designated as the *freezing index F*. This index, expressed in degree-days, is one of the factors of frost penetration (the higher F is, the greater is the frost penetration) (Problems 10.1, 10.2).

10.3 FROST-PENETRATION FACTORS

General

The advance of the frost front in the subsoil (with the exception of the surface temperature) is a function of the soil type, grain-size distribution, freezing index, and of the thermal characteristics of the soil-water system.

Soil Type and Grain-Size Distribution

The classification of soils by their degree of frost susceptibility has been developed by the U. S. Army Corps of Engineers in tabular form (Table 10-1), as well as in chart form (Fig. 10-2).

The tabular form divides soils into four groups designated as F1, F2, F3, F4, where F1 is the least susceptible to frost action and F4 is the opposite extreme (Problems 10.3, 10.4).

The triangular chart maps soils into three domains as acceptable, borderline, and unacceptable in their frost susceptibility (Problem 10.5).

Thermal Characteristics of the Soil-Water System

Specific characteristics of the soil-water system contributing to the frost penetration are listed and defined below:

1. The specific heat, c, is the quantity of heat required to raise the temperature of a unit mass of the material by one degree Fahrenheit. For soil-minerals, -water, and -ice, the value of c may be taken as 0.17, 1.0, and 0.5 respectively. The units are given in Btu per pound per degree Fahrenheit.

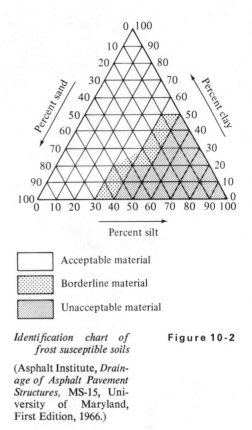

Acceptable material

Borderline material

Unacceptable material

Identification chart of frost susceptible soils   **Figure 10-2**

(Asphalt Institute, *Drainage of Asphalt Pavement Structures*, MS-15, University of Maryland, First Edition, 1966.)

Frost-Susceptible Soils   **TABLE 10-1**

Group	Description
F1	Gravelly soils containing between 3 and 20 percent finer than 0.02 mm by weight.
F2	Sands containing between 3 and 15 percent finer than 0.02 mm by weight.
F3	(a) Gravelly soils containing more than 20 percent finer than 0.02 mm by weight, and sands, except fine silty sands, containing more than 15 per cent finer than 0.02 mm by weight. (b) Clays with plasticity indices of more than 12, and (c) varved clays existing with uniform conditions.
F4	(a) All silts, including sandy silts. (b) Fine silty sands containing more than 15 percent finer than 0.02 mm by weight. (c) Lean clays with plasticity indices of less than 12. (d) Varved clays with nonuniform subgrade.

U.S. Army Corps of Engineers, Engineering and Design, *Pavement Design in Frost Areas,* EM-1110–345–306, *Washington, D.C., 1951.*

2. The *volumetric heat of the soil-water mixture, C,* is the product of the specific heat c and the dry density of the material γ_d. This volume is obviously different for unfrozen and frozen soils and is expressed in units of Btu per cubic foot per degree Fahrenheit.

For unfrozen soils:

$$C_u = \gamma_d c_s + \frac{\gamma_d w c_w}{100} = \left(0.17 + \frac{w}{100}\right) \gamma_d \qquad (10\text{-}1)$$

For frozen soils:

$$C_f = \gamma_d c_s + \frac{\gamma_d w c_i}{100} = \left(0.17 + \frac{w}{200}\right) \gamma_d \qquad (10\text{-}2)$$

where c_s, c_w, c_i are the specific heats of solids, water, ice, respectively, and w is the water content of the system.

3. The *volumetric heat of latent fusion, L,* represents the changes in thermal energy as the soil freezes or thaws at constant temperature. Figure 10-3

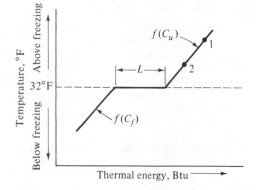

Figure 10-3   *Change in thermal energy with temperature*

shows the change in thermal energy with temperature below and above freezing, interrupted by the latent heat of fusion L of the porewater at freezing temperature (1.0 lb of water yields 143.4 Btu as it freezes). For a soil-water mixture, the latent heat of fusion is

$$L = (1.434\ldots)w\gamma_d \text{ Btu/ft}^3 \qquad (10\text{-}3)$$

4. The *thermal conductivity, k,* is the facility with which the heat is able to travel through a medium. The magnitude of the soil thermal conductivity is controlled by many factors, the most important of which are soil density, soil type, mineral content, organic content, and moisture content.

For bituminous concrete:

$$k \cong 0.84 \text{ Btu/hr/ft/}^\circ\text{F}$$

For portland cement concrete:

$$k \cong 0.54 \text{ Btu/hr/ft/}°\text{F}$$

For soils, since k varies with different types and for frozen and unfrozen conditions, charts (Fig. 10-4) are used to estimate the value of k.

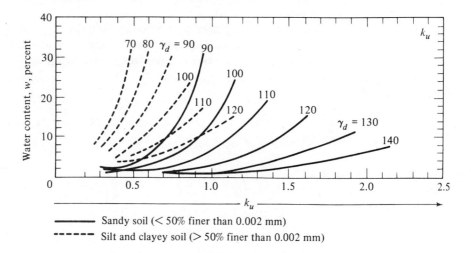

(a)   Unfrozen soil

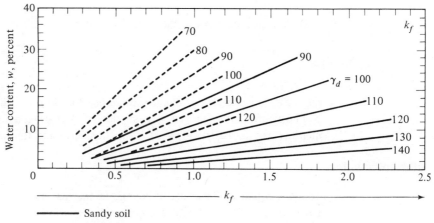

(b)   Frozen soil

Charts of thermal conductivity   **Figure 10-4**

(H. P. Aldrich, "Frost Penetration below Highway and Airfield Pavements," *Highway Research Board Bulletin.*, Vol. 135, 1956, pp. 124–44.)

10.4 FROST-PENETRATION PROCESS

Frost Flow

Whereas the frost (heat) conductivity of soil is called the frost (heat) permeability, the process of frost (heat) transfer in soil (conduction) is called the *penetration*.

The frost (heat) transfer is achieved by radiation, conduction, and/or convection. In fully saturated soils, the heat transfer may be analyzed as the heat conduction, for the radiation and convection are negligible.

Fourier Equation

For fully saturated soils of grain size less than 1.00 mm, the total heat discharge (heat transfer) Q is

$$Q = kiA = \frac{k(T_1 - T_2)A}{l} \tag{10-4}$$

where k = soil heat conductivity,

i = thermal gradient,

T_1, T_2 = temperature at point 1, 2 respectively,

A = transfer area,

l = distance of points 1, 2

By the limiting process, the intensity of transfer q is

$$q = -k\frac{\partial T}{\partial z} \tag{10-5}$$

where $\partial T/\partial z$ is the rate of change in temperature in the z direction.

The Fourier equation (10-4) is similar to Darcy's equation (5-2), for both equations rely on the *principle of conduction* (heat and fluid respectively).

Diffusion Equation

From the equality of the rate of change in thermal energy and of the rate of change in heat flow intensity, the governing equation of the thermal diffusion is

$$a\frac{\partial^2 T}{\partial z^2} = \frac{\partial T}{\partial t} \tag{10-6}$$

where $a = k/C$ is the diffusion constant,

 $T =$ temperature,

 $z =$ penetration depth,

 $t =$ time

The heat-diffusion equation (10-6) is similar to Terzaghi's consolidation equation (6-7), since both equations rely on the *principle of conservation of energy* (Problems 10.6, 10.7).

General Solution

The general solution of (10-6) giving the temperature profile $T = \psi(z, t)$ must satisfy both the differential equation and the boundary conditions associated with the problem. This condition can be met by any of the methods available for the solution of this partial differential equation (Problems 10.8, 10.9, 10.10). More frequently, approximate methods introduced in Section 10.5 are used.

10.5 FROST-PENETRATION ANALYSIS

Approximate Methods

Two approximations for the estimation of depth of frost penetration based on (10-6) are given in this section, (a) the *simplified Stefan method* (1943) and (b) the *modified Berggren method* (1956).

Simplified Stefan Method

The simplest and most commonly used method of frost-penetration analysis is known as the simplified Stefan method.

The initially assumed conditions in this case are that (a) the soil is homogeneous, (b) the volumetric heat of latent fusion, L, is the only heat that must be removed when the soil freezes (i.e., volumetric heat C_u and C_f are neglected), (c) the temperature gradient in the frozen zone is linear, and (d) the pore water is not moving.

The idealized conditions assumed for this formula are shown by Fig. 10-5, where, at the interface between the two zones (any two adjacent zones), the continuity condition must be satisfied. This implies that latent heat supplied by the pore water as it

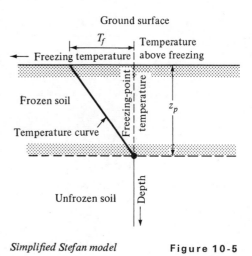

Simplified Stefan model

Figure 10-5

freezes to a depth of dz in time dt equals the rate at which the heat is conducted to the ground surface.

This equality eventually yields the estimated depth equation

$$z_p = \sqrt{\frac{48k_f F}{L}} \tag{10-7}$$

where z_p = frost-penetration depth,

$\quad\quad k_f$ = heat conductivity of frozen soil (Fig. 10-4b),

$\quad\quad F$ = freezing index,

$\quad\quad L$ = latent heat of fusion (10-3)

The derivation, limitation, and use of this approximate formula are given in Problems 10.11, 10.12, and 10.13. It must be stressed here that (10-7) overestimates z_p.

Modified Berggren Method

The rigorous solution of the diffusion equation (10-6) developed by Berggren (1956), and modified by Aldrich and Paynter for the conditions of Fig. 10-6, yields the depth of frost penetration,

$$z_p = \lambda\sqrt{\frac{48k_f F}{L}} \tag{10-8}$$

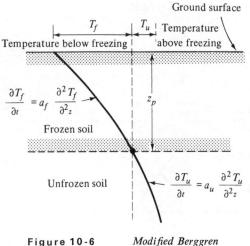

where k_f, F, L are defined in (10-7) and λ is a dimensionless correction factor.

The correction factor λ (always less than unity) depends on the thermal properties of the unfrozen and frozen soil and is a function of two parameters, the thermal ratio α and the fusion parameter μ (Fig. 10-7). They are

$$\alpha = \frac{T_u}{T_f} \quad\text{or}\quad = \frac{T_u t}{F} \tag{10-9}$$

$$\mu = \frac{C_f}{L} T_f \cong \frac{C_f F}{L t} \tag{10-10}$$

Figure 10-6 *Modified Berggren model*

where T_f, T_u are the temperatures shown in Fig. 10-6, t is the duration of the freezing period, and C_f, F, L are defined in (10-2) and (10-7).

The limitation and use of (10-8) are given in Problems 10.14 and 10.15. Finally, the extension of (10-8) to multilayer systems is shown in Problems 10.16, 10.17, and 10.18.

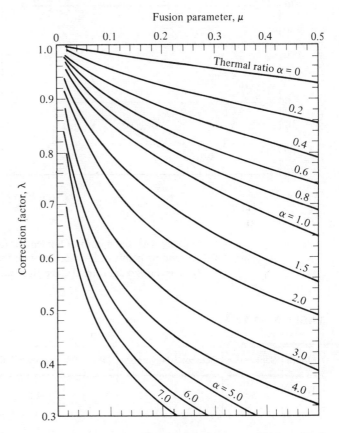

Fusion parameter, μ

Correction factor

Correction factor, λ

Thermal ratio $\alpha = 0$

Figure 10-7

(H. P. Aldrich, "Frost Penetration below Highway and Airfield Pavements," *Highway Research Board Bulletin*, Vol. 36, 1956, pp. 124–
42.)

SOLVED PROBLEMS

Freezing Index

10.1   Compute the freezing index F from the mean monthly temperature data
recorded at a given location and listed Table P-10.1a. Use the trapezoidal formula.

By definition, the freezing index F is the area enclosed by the annual temperature   *Solution*
profile below the time axis. The evaluation of this area may be accomplished by
any of the methods available for the quadrature calculation.

   Regardless of the method used, the vertical coordinates of the temperature annual profile must be computed first as the difference between the mean monthly
temperature and the freezing temperature (Table P-10.1b, column 2).

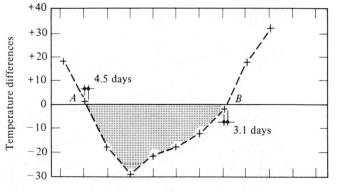

Figure P-10.1

Starting with the month of September (any month), the profile is shown in Fig. P-10.1, and its points of intersection with the time axis are identified as A and B. By the trapezoidal formula, the freezing index is then

TABLE P-10.1a

Month	September	October	November	December	January	February	March	April	May	June
Temp. °F	50	35	15	3	10	15	20	30	50	65

TABLE P-10.1b

Month 1	Difference Between Mean Monthly Temperature and Freezing Temperature 2	Days/ Month 3	Degree-days 4	Cumulative Degree-days 5
Sept.	$50 - 32 = +18$	30	$= +540$	$+ 540$
Oct.	$35 - 32 = + 3$	31	$= + 93$	$+ 633$
Nov.	$15 - 32 = -17$	30	$= -510$	$+ 123$
Dec.	$3 - 32 = -29$	31	$= -899$	$- 776$
Jan.	$10 - 32 = -22$	31	$= -682$	-1458
Feb.	$15 - 32 = -17$	28	$= -476$	-1934
Mar.	$20 - 32 = -12$	31	$= -372$	-2306
Apr.	$30 - 32 = - 2$	30	$= -060$	-2366
May	$50 - 32 = +18$	31	$= +558$	-1808
June	$65 - 32 = +33$	30	$= +990$	$- 818$

$$F = -(17)\left(\frac{25.5}{2}\right) - (17+29)\left(\frac{31}{2}\right) - (29+22)\left(\frac{31}{2}\right)$$

$$- (22+17)\left(\frac{28}{2}\right) - (17+12)\left(\frac{31}{2}\right) - (12+2)\left(\frac{30}{2}\right) - (2)\left(\frac{3}{2}\right)$$

$$\cong 2962 \text{ degree-days}$$

Because of the shape of the wave, a very close result can be obtained by summing the negative numbers in column 4 of Table P-10.1b, that is,

$$F \cong -510 - 899 - 682 - 476 - 372 - 60 = 2999 \text{ degree-days}$$

which is obviously the application of the rectangular formula.

10.2   Using the cumulative degree-day curve of Fig. P-10.2, compute the freezing index of the temperature histograph given in Table P-10.1b.

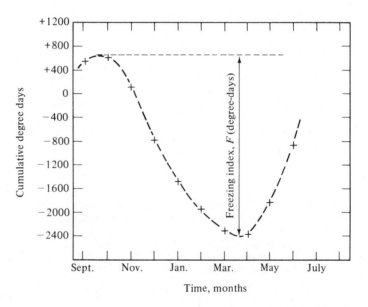

Figure P-10.2

The cumulative degree-days are calculated from the data given in Table P-10.1a.   *Solution* The procedure of calculation is shown in Table P-10.1b.

Then the cumulative degree-days are plotted versus the months, as shown in Fig. P-10.2.

The distance between the upper and lower inflection points of the curve represents the surface freezing index F, which in this case is

$$F \cong 3680 \text{ degree-days}$$

This result is usually higher than the result obtained by means of the trapezoidal formula shown in Problem 10.1.

Frost Penetration

10.3 Using the U. S. Army Corps of Engineers classification (Table 1-10), determine the degree of frost susceptibility for the two soils, the properties of which are given in Table P-10.3.

Solution According to Table 10-1, soil 1 is classified as F2; soil 2 is classified as F4. Consequently, soil 2 is more frost susceptible than soil 1.

TABLE P-10.3

1. Sieve Analysis									
Sieve size	4	10	20	40	100	200	270	0.005 mm	0.001 mm
Soil 1, percent passing	95	98.1	74.8	58.2	40.1	3.8	1.2	–	–
Soil 2, percent passing	98	94		80		57	50	20	15

2. Index Properties
Soil 1 is nonplastic Soil 2 has a LL = 47 % and PL = 35 %

3. Unified System Classification
Soil 1 has only 3.8 % passing No. 200 sieve; thus it belongs to the coarse-grained group. But it has greater percentage of coarse fraction passing No. 4 and less than 5 % passing No. 200 sieves. Therefore the material is classified as sand, or SP. Soil 2 has more than 50 % passing No. 200 sieve; thus it belongs to the fine-grained group. Using Fig. 2-3 and introducing the index data given in the problem, LL = 47 % and PI – 47 – 35 = 12 % gives the soil class as ML or silt with low compressibility.

10.4 From the information of Table P-10.3, judge the frost penetration in soil 1 and soil 2.

Answer As indicated in the charts of Fig. 10-4, the frost penetration will be greater in soil 1 because sands and granular soils, in general, have higher thermal conductivity than finer soils.

10.5 Three fine-grained soils have the makeup indicated in Table P-10.5. Using the Asphalt Institute triangular chart (Fig. 10-2), classify these soils according to their frost susceptibility.

	% Sand	% Silt	% Clay
Soil 1	50	33	17
Soil 2	40	20	40
Soil 3	10	80	10

By plotting the percentages of sand, silt, and clay on the sides of the triangular   *Solution*
chart (Fig. 10-2), the point on the chart representing the susceptibility of the soil is
determined. Using this method for the three soils given in the problem yields

Soil 1: borderline material

Soil 2: acceptable material

Soil 3: unacceptable material

Frost-Penetration Process

10.6   Consider a fully saturated soil of grain size less than 1.00 mm and an idealized
heat flow in this soil between two of its points, 1 and 2. If the distance of these
points is l, the heat-flow transfer area is A, the temperatures at 1 and 2 are T_1 and
T_2, respectively, and the soil heat conductivity is a known value k, derive the heat
discharge equation (10-4).

Similar to the fluid-flow discharge equation called Darcy's law (1), (5-2), the heat   *Solution*
discharge equation is a result of experiments and observations.
    The simplest case of conduction investigated by Fourier (1807) is that of single
panel wall of Fig. P-10.6 through which heat is transferred normal to the wall
surface.

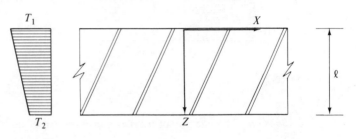

Figure P-10.6

If the wall of thickness l is assumed to be infinite in the $\pm x$ direction (to eliminate the effect of end condition) and the face temperature are measured values $T_1 > T_2$, then the thermal gradient, defined as the temperature change per unit length, is

$$i = \frac{T_1 - T_2}{l} \tag{a}$$

where T_1, T_2 are assumed to remain constant during the experiment.

Furthermore, if the thermal conductivity k of the wall material is a known value, the intensity of transfer becomes a product ki; that is,

$$q = ki = \frac{k(T_1 - T_2)}{l} \tag{b}$$

and is defined as the heat discharge per unit area.

The total heat discharge through the egress area A is then

$$Q = kiA = \frac{k(T_1 - T_2)A}{l} \tag{c}$$

which is the desired relationship (10-4).

This simple deduction developed and experimentally verified by Fourier is directly applicable in engineering soil mechanics.

10.7   Since layered (stratified) soils frequently occur in the subgrade, extend equation (c) of Problem 10.6 to the case of soil consisting of m layers parallel to the XY plane. Assume the thickness of these layers as $l_1, l_2, \ldots, l_m$, their thermal conductivity as $k_1, k_2, \ldots, k_m$, and the temperatures at their boundary as $T_1, T_2, \ldots, T_m$ (Fig. P-10.7).

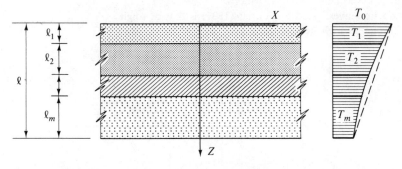

Figure P-10.7

Solution   The intensity of transfer q given by (b) of Problem 10.6 can be written as

$$q = \frac{\Delta T}{R} \tag{a}$$

where

$$R = \frac{l}{k} \tag{b}$$

is denoted as the thermal resistance.

Since (a) of this problem is similar to Ohm's law (1820), which states that

$$I = \frac{E}{R} = \frac{E}{R_1 + R_2 + \cdots + R_m} \tag{c}$$

where $I =$ discharge of electric current,

$E =$ electromotive force,

$R =$ conductor's resistance

From the definition of composite resistance in (c), the composite resistance in (a) can be expressed as

$$R = \frac{l_1}{k_1} + \frac{l_2}{k_2} + \cdots + \frac{l_m}{k_m} \tag{d}$$

With (d), equation (a) becomes

$$q = \frac{T_0 - T_m}{l_1/k_1 + l_2/k_2 + \cdots + l_m/k_m} \tag{e}$$

and

$$Q = \frac{(T_0 - T_m)A}{\sum\limits_{j=1}^{m} l_j/k_j} \tag{f}$$

which is the desired discharge equation for the soil system of Fig. P-10.7.

Diffusion Equation

10.8   Consider a homogeneous, isotropic, and continuous semi-infinite soil system of known thermal conductivity k. Assuming the temperature above or below the freezing, derive the temperature-depth-profile differential equation designated as the diffusion equation (10-6).

With the assumption of one-directional heat flow in the z direction, the conser- *Solution* vation of thermal energy required that the rate of change in thermal energy $\partial u/\partial t$ equal the rate of change in heat flow intensity $\partial q/\partial z$; that is,

$$\frac{\partial u}{\partial t} = -\frac{\partial q}{\partial z} \tag{a}$$

The time rate of thermal energy is given experimentally as

$$\frac{\partial u}{\partial t} = \frac{\partial T}{\partial t} C \tag{b}$$

where C is the volumetric heat of the soil-water mixture given by (10-1) or (10-2).

The position derivative of q, derived by differentiating (10-5), is

$$\frac{\partial q}{\partial z} = -\frac{k\partial^2 T}{\partial z^2} \tag{c}$$

Then (a) in terms of (b) and (c) becomes

$$a\frac{\partial^2 T}{\partial z^2} = \frac{\partial T}{\partial t} \tag{d}$$

where $a = k/C$.

The resulting differential equation (d) is the desired relation (10-6), which can be used for the estimation of the frost penetration.

10.9   Using the method of separation of variables, find the general integral of the diffusion equation (10-6).

Solution   In order to find the integral of (10-6), a product function

$$T(t, z) = Y(t)Z(z) \tag{a}$$

is assumed to be the solution.

Then (10-6), in terms of partial derivatives of (a), becomes

$$\frac{aZ''}{Z} = \frac{\dot{Y}}{Y} \tag{b}$$

where

$$Z'' = \frac{\partial^2 Z(z)}{\partial z^2} \quad \text{and} \quad \dot{Y} = \frac{\partial Y(t)}{\partial t}$$

Since the left side in (b) is a function of z and the right side is a function of t, each side must be equal to the same arbitrary constant (say $\pm\lambda^2$).

$$\frac{aZ''}{z} = \mp\lambda^2 \quad \text{and} \quad \frac{\dot{Y}}{Y} = \mp\lambda^2 \tag{c, d}$$

where the minus sign is taken for the decrease in temperature with time and the plus sign is introduced for the opposite case.

Hence

$$Z'' + \frac{\lambda^2}{a}Z = 0 \qquad \dot{Y} + \lambda^2 Y = 0 \tag{e, f}$$

the solutions of which are

$$Z(z) = C_1 \cos \frac{\lambda z}{\sqrt{a}} + C_2 \sin \frac{\lambda z}{\sqrt{a}} \qquad Y(t) = C_3 e^{-\lambda^2 t} \qquad \text{(g, h)}$$

The product of (g, h) is then the general integral of (10-6), given as

$$T(t, z) = \left[C_1 \cos \left(\frac{\lambda z}{\sqrt{a}} \right) + C_2 \sin \left(\frac{\lambda z}{\sqrt{a}} \right) \right] C_3 e^{-\lambda^2 t} \qquad \text{(i)}$$

where C_1, C_2, C_3 are the constants of integration and λ is a characteristic value taken as

$$\lambda = m\pi, \qquad m = 0, 1, 2, \ldots \qquad \text{(j)}$$

10.10 Show the evaluation of C_1, C_2, C_3 in equation (i) of Problem 10.9 for a typical soil system.

For the evaluation of C_1, C_2, C_3, the annual temperature profile must be known at   *Solution*
$t = t, z = 0$, and $t = t, z = \sqrt{a}/2$.
  For $t = t, z = 0$,

$$C_1 C_3 = T(t, 0) e^{-m^2 \pi^2 t} \qquad \text{(a)}$$

  For $t = t, z = \sqrt{a}/2$,

$$C_2 C_3 = T\left(t, \frac{\sqrt{a}}{2}\right) e^{-m^2 \pi^2 t} \qquad \text{(b)}$$

  With (a, b), the general integral becomes

$$T(t, z) = \sum_{m=0}^{m=\infty} \left[T(t, 0) \cos \left(\frac{m\pi z}{\sqrt{a}} \right) + T\left(t, \frac{\sqrt{a}}{2}\right) \sin \left(\frac{m\pi z}{\sqrt{a}} \right) \right] e^{-m^2 \pi^2 t}$$

where $\sqrt{a}/2$ is the characteristic depth.

Approximate Methods

10.11 Derive the Stefan formula (10-7) for the frost-depth penetration in a homogeneous soil system. Assume the following given values: $F =$ freezing index, $T_f =$ surface temperature below freezing, $k_f =$ the heat conductivity of frozen soil, $L =$ the latent heat of fusion.

Figure 10-5 shows the idealized conditions assumed for the Stefan model. At   *Solution*
the interface between the two zones, the equation of continuity must be satisfied; that is, the latent heat supplied by the pore-soil moisture as it freezes to a depth Δz in time Δt must equal the rate at which heat is conducted to ground surface.
  Thus

$$L \frac{dz}{dt} = \frac{k_f T_f}{z} \qquad \text{(a)}$$

where z is the position coordinate (depth below surface).

The integration of (a) in limits $t = 0, t = t$ and $z = 0, z = z_p$ yields

$$\frac{k_f}{L} \int_0^t T_f \, dt = \int_0^{z_p} z \, dz \tag{b}$$

where

$$\int_0^t T_f \, dt \cong F = \text{the equivalent surface freezing index in degree-hours.}$$

The evaluation of $\int_0^t T_f \, dt$ in degree-days and the evaluation of the right side in given limits gives

$$z_p = \sqrt{\frac{48k_f F}{L}} \tag{c}$$

which is (10-7).

10.12   Discuss the limitations of the Stefan approximate depth equation (10-7).

Answer   The limitations of this simplified relationship (10-7) evolve from the following:

1. the assumed condition of linear variation of temperature with depth.
2. the assumed condition of constant temperature below the frozen zone and at the surface.
3. neglecting the volumetric heat factors of both frozen and unfrozen soils.

Thus the use of this approximate formula will result in an overestimated depth of frost penetration.

The extent of overestimation depends on such factors as moisture content of the soil and climatic conditions at the site, factors that will affect the relative values of the neglected volumetric heat to that of the latent heat.

10.13   Using the Stefan formula (10-7), calculate the depth of frost penetration in a uniform silty-loam soil that has the following characteristics: moisture content, $w = 20\%$, dry density $\gamma_d = 100 \, \text{lb/ft}^3$. The weather data yielded the following information: mean annual temperature = 42 °F, duration of freezing period = 100 days, and the freezing index = 2000 degree-days.

Solution   To use the Stefan formula (10-7), the following parameters must be calculated first. By (10-3), the volumetric heat of latent fusion L is

$$L = (1.434)w\gamma_d = (1.434)(20)(100) = 2868 \, \text{Btu/ft}^3$$

From Fig. 10-4, the value of the thermal conductivity of frozen soil k_f is 1.03 Btu/hr/ft/°F. By (10-7), the depth of frost penetration is

$$z_p = \sqrt{\frac{48k_f F}{L}} = \sqrt{\frac{(48)(1.03)(2000)}{2868}} = 5.87 \, \text{ft}$$

10.14   Using the Berggren formula (10-8), calculate the depth of frost penetration in the soil whose characteristics are given in Problem 10.13.

To use the Berggren formula (10-8), the following parameters must be calculated   *Solution* first. By (10-1), the volumetric heat of unfrozen soil C_u is

$$C_u = \left(0.17 + \frac{w}{100}\right)\gamma_d = (0.17 + 0.2)(100) = 37.0 \text{ Btu/ft}^3/°\text{F}$$

By (10-2), the volumetric heat of frozen soil C_f is

$$C_f = \left(0.17 + \frac{w}{200}\right)\gamma_d = (0.17 + 0.1)(100) = 27.0 \text{ Btu/ft}^3/°\text{F}$$

Hence

$$C_{\text{eff}} = \frac{C_f + C_u}{2} = \frac{37.0 + 27.0}{2} = 32.0 \text{ Btu/ft}^3/°\text{F}$$

Thermal conductivity of frozen soil k_f is determined in Problem 10.13 and has a value of $k_f = 1.03$ Btu/hr/ft^3/°F. From Fig. 10-4a, the value of the thermal conductivity of unfrozen soil $k_u = 0.80$ Btu/hr/ft^3/°F. Hence

$$k_{\text{eff}} = \frac{k_u + k_f}{2} = \frac{0.8 + 1.03}{2} = 0.915 \text{ Btu/hr/ft/}°\text{F}$$

By (10-9), the thermal ratio is

$$\alpha = \frac{T_f t}{F} = \frac{(42 - 32)(100)}{2000} = 0.5$$

By (10-10), the fusion parameter is

$$\mu = \frac{C_{\text{eff}} F}{Lt} = \frac{(32)(2000)}{(2868)(100)} = 0.223$$

By introducing values of α and μ in Fig. 10-7, the correction factor λ is found to be $= 0.853$. The volumetric heat of latent fusion L (calculated in Problem 10.13) is

$$L = 2868 \text{ Btu/ft}^3 \tag{a}$$

Using the modified Berggren formula (10-8), the depth of frost penetration is

$$z_p = \lambda\sqrt{\frac{48k_f F}{L}} = 0.853\sqrt{\frac{(48)(0.915)(2000)}{2868}} = 4.72 \text{ ft} \tag{b}$$

10.15   Compare the values of z_p obtained by the Stefan approximate formula (10-7) in Problem 10.13 and that obtained by the modified Berggren formula (10-8) in Problem 10.14. Comment on the results of your comparison.

Answer It is obvious from the results obtained in Problems 10.13 and 10.14 that the Stefan formula (10-7) yields a higher value for the estimated depth of frost penetration z_p than the Berggren formula (10-8).

However, it may be also noted that the estimated depth from both formulas will be approximately the same when the value of the correction value λ approaches unity. The value of λ will approach unity under the following conditions:

1. in soils having a high moisture content.
2. in areas where the mean annual surface temperature equals that of the freezing temperature of the soil-water system.

10.16 Show how the Berggren formula (10-8) can be used to determine the depth of frost penetration in a nonuniform (stratified) soil system.

Answer The modified Berggren formula, with some modification, can be used for the determination of z_p in a stratified soil. The general formula in this case is

$$z_p = \lambda \sqrt{\frac{48F}{(L/k)_{\text{eff}}}} \qquad (a)$$

where

$$\left(\frac{L}{k}\right)_{\text{eff}} = \frac{2}{z_e^2}\left\{\frac{d_1}{k_1}\left[\left(\frac{L_1 d_1}{2}\right) + L_2 d_2 + \cdots + L_n d_n\right] + \frac{d_2}{k_2}\left[\left(\frac{L_2 d_2}{2}\right)\right.\right.$$
$$\left.\left. + L_3 d_3 + \cdots + L_n d_n\right] + \cdots + \frac{d_n}{k_n}\left(\frac{L_n d_n}{2}\right)\right\} \qquad (b)$$

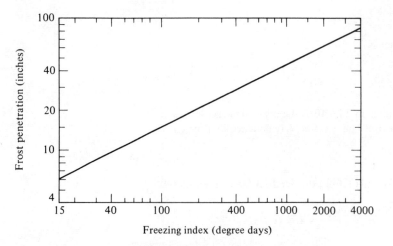

Figure P-10.16   (Corps of Engineers, *Report of Pavement Surface Temperature Transfer Study,* Frost Effects Laboratory, New England Division, Boston, Mass., 1950.)

where d = depth,

   L = latent heat of fusion,

   k = thermal conductivity,

and subscripts 1, 2, 3, etc., refer to the different layers of the stratified system. Furthermore z_e is the preliminary rough estimate of the depth of the frost penetration in the whole system. This value of z_e is determined from the Corps of Engineers chart (Fig. P-10.16), which was originally developed as an empirical relationship to estimate the depth of frost penetration in a well-drained granular material on the basis of the freezing index F (see Problem 10.17).

10.17   Estimate the approximate depth of frost penetration z_p in a granular soil and with the freezing index equal to that determined in Problem 10.1.

Since the material is a granular, nonfrost-susceptible soil, the Corps of Engineers *Solution* chart (Fig. P-10.16) is used. From this chart and for a freezing index $F = 2999$ degree-days, the approximate depth of frost penetration $z_e = 74$ in.

10.18   Determine the depth of frost penetration in a layered system of flexible pavement. Thickness and characteristics of each layer in the system are listed in Table 10.18a. The freezing index F is 1500 degree-days; mean annual temperature is 35 °F.

TABLE P-10.18a

Material	Thickness (in.)	Water Content (w %)	γ_d (lb/ft³)
Bituminous concrete surface	3.0	0	145
Base	6.0	6.50	110
Subbase	21.0	8.50	90
Clay subgrade	–	29.0	90

The best way to approach the solution of such a problem is to organize the solution *Solution* into steps as follows:

Step 1.   The freezing index $F = 1500$ degree-days (If not given in the problem, it should be calculated from weather data; see Problems 10.1, 10.2)

Step 2.   Duration of the freezing period $t = 150$ days and mean annual temperature = 35 °F.
   Thus

$$T_u = 35 - 32 = 3 \text{ °F}$$

Step 3.   Determination of the thermal properties of each layer:
   (a)   The bituminous concrete surface:

By (10-3), the volumetric heat of fusion L is

$$L = (1.434)w\gamma_d = (1.434)(0)(145) = 0 \text{ Btu/ft}^3$$

The volumetric heat C for bituminous concrete is $= 28 \text{ Btu/ft}^3/°F$
The thermal conductivity k for bituminous concrete is $0.84 \text{ Btu/ft}^3/\text{hr}/°F$

(b) The base course:
By (10-3), the volumetric heat of fusion L is

$$L = (1.434)w\gamma_d = (1.434)(6.5)(110) = 1025 \text{ Btu/ft}^3$$

By (10-1), the volumetric heat of unfrozen soil is

$$C_u = \left(0.17 + \frac{w}{100}\right)\gamma_d = \left(0.17 + \frac{6.5}{100}\right) = 25.9 \text{ Btu/ft}^3/°F$$

By (10-2), the volumetric heat of frozen soil is

$$C_f = \left(0.17 + \frac{w}{200}\right)\gamma_d = \left(0.17 + \frac{6.5}{200}\right) = 22.3 \text{ Btu/ft}^3/°F$$

Thus,

$$C_{av} = \frac{C_u + C_f}{2} = \frac{25.9 + 22.3}{2} = 24.10 \text{ Btu/ft}^3/°F$$

Thermal conductivity values for frozen and unfrozen conditions are determined from Fig. 10-4a and b (using the sandy soil curves) and for values of $w = 6.5 \%$ and $\gamma_d = 110 \text{ lb/ft}^3$. The resulting values of k will be

$$k_f = 0.92 \text{ Btu/ft}^3/\text{hr}/°F$$
$$k_u = 1.0 \text{ Btu/ft}^3/\text{hr}/°F$$

(c) Similarly, the thermal properties of the subbase and subgrade are calculated. To facilitate subsequent steps of calculations, results are tabulated as shown in Table P-10.18b.

Step 4. Computation of $(L/k)_{\text{eff}}$:

It is necessary at this step to estimate approximately the total depth of frost penetration into the combined system, as well as the approximate depth of subgrade material included in the total depth of penetration. We can do so by introducing the value of $F = 1500$ degree-days (from Step 1) into Fig. P-10.16 and reading off the value of z_e, which in this case is 55 in. or 4.6 ft. Thus the approximate depth of frozen subgrade is

$$d_4 = [55 - (3 + b + 21)] = 25 \text{ in.}$$

Then by using equation (b) of Problem 10.16, we obtain

Property	Surface	Layered System		
		Base	Subbase	Subgrade
L	0	1025	1279	3742
C_f	–	22.30	22.30	28.40
C_u	–	25.90	26.80	41.40
C_{av}	28	24.10	42.55	34.90
k_f	–	0.92	0.96	1.20
k_n	–	1.00	0.98	0.73
k_{av}	0.84	0.96	0.97	0.97

$$\left(\frac{L}{k}\right)_{eff} = \frac{2}{(4.6)^2}\left\{\frac{0.25}{0.84}\left[\frac{(0)(0.25)}{2} + (1025)(0.5) + (1279)(1.75)\right.\right.$$

$$\left. + (3742)(2.083)\right] + \frac{0.5}{0.96}\left[\frac{(1025)(0.5)}{2} + (1279)(1.75)\right.$$

$$\left. + (3742)(2.083)\right] + \frac{2.083}{0.965}\left[\frac{(3742)(2.083)}{2}\right]\right\} = 3118$$

Step 5. Computation of the weighted (average) values of C and L:

To determine the value of the fusion parameter μ (Step 6), weighted (average) values of C and L for the combined system are introduced into the formula (10-10). Thus they must be calculated first as follows:

$$C_{wt} = \frac{(28)(0.25) + (24.10)(0.5) + (24.55)(1.75) + (34.9)(2.083)}{4.60}$$

$$= 29.28 \text{ Btu/(ft}^3)(°F)$$

$$L_{wt} = \frac{(0)(0.25) + (1025)(0.5) + (1279)(1.75) + (3742)(2.083)}{4.60}$$

$$= 2292 \text{ Btu/ft}^3$$

Step 6. Computation of effective values of α and μ:

By (10-9), the thermal ratio is

$$\alpha = \frac{T_u t}{F} = \frac{(3)(150)}{1500} = 0.300$$

By (10-10), the fusion parameter (modified for stratified system) is

309

$$\mu = \frac{(C_{wt})(F)}{(L_{wt})(t)} = \frac{(29.28)(1500)}{(2292)(150)} = 0.1277$$

where C_{wt} and L_{wt} are the weighted average of C and L calculated in Step 5.

TABLE P-10.18c *Frost Penetration—Layered System, Nonuniform Soil*

Pavement Type: Highway Airfield Mean (ave) Annual Temp. _____ °F $\quad L = 1.434(w)(\gamma_d) \quad C_f = \gamma_d\left(0.17 + \frac{0.5w}{100}\right)$

$F =$ _____ Degree-days $\qquad t =$ _____ Days $\qquad C_u = \gamma_d\left(0.17 + \frac{w}{100}\right)$

Layer	Thick-ness (ft)	Density (lb/ft³)	Moisture Content %	Thermal Conduc-tivity	Average	$\frac{z}{K}$	Latent Heat	Lz	Heat Capacity	Average C	Cz
1	$z_1=$	$\gamma_1=$ Type $=$	$w_1=$	$K_{u_1}=$ $K_{f_1}=$	$K_1=$	$\frac{z_1}{K_1}=$	$L_1=$	$L_1z_1=$	$C_{u_1}=$ $C_{f_1}=$	$C_1=$	$C_1z_1=$
2	$z_2=$	$\gamma_2=$ Type $=$	$w_2=$	$K_{u_2}=$ $K_{f_2}=$	$K_2=$	$\frac{z_2}{K_2}=$	$L_2=$	$L_2z_2=$	$C_{u_2}=$ $C_{f_2}=$	$C_2=$	$C_2z_2=$
3	$z_3=$	$\gamma_3=$ Type $=$	$w_3=$	$K_{u_3}=$ $K_{f_3}=$	$K_3=$	$\frac{z_3}{K_3}=$	$L_3=$	$L_3z_3=$	$C_{u_3}=$ $C_{f_3}=$	$C_3=$	$C_3z_3=$
4	$z_4=$	$\gamma_4=$ Type $=$	$w_4=$	$K_{u_4}=$ $K_{f_4}=$	$K_4=$	$\frac{z_4}{K_4}=$	$L_4=$	$L_4z_4=$	$C_{u_4}=$ $C_{f_4}=$	$C_4=$	$C_4z_4=$

$z_m = [z_{p(est)} - (z_1 + z_2 + z_3 + z_4 + \cdots + z_{(n+1)})]$ $\qquad$ Total $Lz =$ _____ $\qquad$ Total $Cz =$ _____

$z_{p(est)} =$ _____ (from Fig. P-10.16) $\qquad T_u =$ mean (ave) tem. (32 °F) $\qquad L_{wt} = \dfrac{\text{total } Lz}{[z_{p(est)}]} =$ _____ $=$ _____

$[z_{p(est)}]^2 =$ _____ $\qquad\qquad T_u =$ _____ °F $\qquad\qquad C_{wt} = \dfrac{\text{total } Cz}{[z_{p(est)}]} =$ _____ $=$ _____

	$L_1z_1=$	$L_2z_2=$	$L_3z_3=$	$L_4=$	Subtotal
$\frac{z_1}{K_1}=$	$\frac{2}{2}=$				
	$\frac{z_2}{K_2}=$	$\frac{2}{2}=$			
		$\frac{z_3}{K_3}=$	$\frac{2}{2}=$		
			$\frac{z_4}{K_4}=$	$\frac{2}{2}=$	

$\gamma = \dfrac{T_u t}{F} =$ _____ $=$ _____

$\mu = \dfrac{C_{wt}(F)}{L_{wt}(T)} =$ _____ $=$ _____

$\lambda =$ (from Fig. 10.7) $=$ _____

$Z_p = \lambda\sqrt{\dfrac{48\,F}{(L/K)_{\text{eff}}}} =$ _____

$z_p =$ _____

$\dfrac{L}{K_{\text{eff}}} = \dfrac{2(\text{total } A)}{[z_{p(est)}]^2}$

$\dfrac{L}{K_{\text{eff}}} =$ _____ $=$ _____ $\qquad\qquad$ Total $A =$ _____

Step 7. Computation of the correction factor λ:

Introducing the values of α and μ (calculated in Step 6) into Fig. 10-7, the value of λ is found to be $= 0.93$.

Step 8. Determination of the depth of frost penetration z_p in the combined system:

By equation (a) of Problem 10.16, the depth of frost penetration is

$$z_p = \lambda\sqrt{\frac{48F}{(L/k)_{\text{eff}}}} = 0.93\sqrt{\frac{(48)(1500)}{3118}} = 4.47 \text{ ft}$$

Since, in this case, the calculated value of z_p is $= 4.47$ ft, which is very close to the initially assumed value of $z_p = 4.58$ ft, no additional readjustment of the calculated value is necessary. If the values are appreciably different, then repeat steps 4 to 8, using a value of $z_p =$ average of the initially assumed value of Z_e and the calculated value z_p. In general, not more than two trials will be necessary.

The solution of such a problem, as can be seen from the preceding calculation, is usually lengthy and elaborate. However, the organization of the solution of this problem in a tabular form may greatly simplify the computational processes. Table P-10.18c is suggested as a convenient arrangement of the calculations for this problem. All terms used in this table are the same as those in Problems 10.17 and 10.18. The student may use a copy of this table and fill in the blank spaces with the computations in Problem 10.18.

SUPPLEMENTARY PROBLEMS

Freezing Index

10.19 The following average monthly temperatures were obtained at the site of a paving project. From this data, calculate the surface freezing index, using the trapizoidal formula and the histograph method.

Month	September	October	November	December	January	February	March	April	May
Temperature °F	60	45	35	2	22	26	30	42	50

Freezing index $F \cong 1490$ degree-days   *Answer*

Frost Penetration

10.20 Determine the depth of freezing in a silty-clay soil that has a dry density of 100 lb/ft³ and a moisture content of 15 percent. Freezing index calculated from weather records is 1000 degree-days, duration of freezing period is 1000 days, and the mean annual air temperature is 45 °F.

Depth of frost penetration $z_p = 3.22$ ft.   *Answer*

10.21 Determine the depth of frost penetration in a sandy soil that has a specific gravity of 2.65. The weather data indicate a mean annual surface temperature of 42 °F, and the duration of the freezing period is 100 days. Freezing index is 1000 degree-days.

Answer Depth of frost penetration $z_p = 3.91$ ft.

10.22 What three basic factors are necessary for the occurrence of a damaging effect to pavement structures and other surface footings from frost action ?

Answer Refer to Section 10.1, page 286.

Frost Penetration Through Stratified Soil

10.23 A pavement consists of 3 in. of a bituminous concrete surface, 12 in. of sandy-gravelly base having a dry density of 125 lb/ft³ and moisture content of 10 percent, over a silty-clay subgrade that has a moisture content of 25 percent and natural wet density of 125 lb/ft³. The mean annual temperature is 70 °F, the freezing index is 1200 degree-days, and the duration of freezing period is 120 days. Calculate the depth of frost penetration in this structure.

Answer Depth of frost penetration $z_p \cong 2.63$ ft.

references
and
selected
bibliography

CHAPTER 1

Morphology

1.01 Casagrande, A., Research of the Atterberg Limits of Soils, Public Roads, Vol. 13, No. 8, Oct. 1932.

1.02 Committee Report, Glossary of Terms and Definitions in Soil Mechanics, Proc. ASCE, Journal of The Soil Mechanics and Foundation, Vol. 84, SM 4, Oct. 1958.

1.03 Deresiewicz, H., Mechanics of Granular Matter, Advances in Applied Mechanics, Vol. 5, Academic Press, New York, 1958, p. 233.

1.04 Dunham, C. W., Foundations of Structures, McGraw-Hill, New York, 1950, pp. 7-23.

1.05 Hewes, L. I., and C. H. Oglesby, Highway Engineering, Wiley, New York, 1954, p. 399.

1.06 Hough, B. K., Basic Soil Engineering, Ronald Press, New York, 1957, pp. 30-31.

1.07 Jumikis, A. R., Soil Mechanics, Van Nostrand, New York, 1962, pp. 90-92.

1.08 Krynine, D. P., and William Judd, Principles of Engineering Geology and Geotechnics, McGraw-Hill, New York, 1957, pp. 81-125.

1.09 Means, R. E., and J. V. Parcher, Physical Properties of Soils, Charles Merrill, Columbus, Ohio, 1963, pp. 24-28.

1.10 Peck, R., W. Hanson, T. H. Thornburn, Foundation Engineering, Wiley, New York, 1953, pp. 3-34.

1.11 Sowers, G. B., and G. F. Sowers, Introductory Soil Mechanics and Foundations, 2nd Ed., MacMillan, New York, 1961, pp. 1-28.

1.12 Terzaghi, K., and Ralph Peck, Soil Mechanics in Engineering Practice, Wiley, 1948, pp. 3-30.

1.13 The Asphalt Institute, Soils Manual, Manual Series No. 10, College Park, Maryland, May 1964, p. 23.

1.14 Thornbury, T. W., Principles of Geomorphology, Wiley, New York, 1954, pp. 68-97.

CHAPTER 2

Classification

2.01 Allen, H., Classification of Soils and Control Procedures Used in Construction of Embankments, Public Roads, Vol. 22, No. 12, 1942, pp. 263-282.

2.02 Casagrande, A., Classification and Identification of Soils, Proc. ASCE, Vol. 74, No. 3, 1948, pp. 407-409; Discussion by K. S. Lane, pp. 121-131; Discussion by D. M. Burmister, pp. 409-415.

2.03 Casagrande, A., Classification and Identification of Soils, Trans., ASCE Vol. 113, 1948, p. 901.

2.04 Hewes, L., and C. H. Oglesby, Highway Eng., Wiley, New York, 1954, pp. 420-430.

2.05 Highway Research Board, Soil Mapping Methods and Applications, H. R. B., Vol. 41, Bull. No. 299, Wash., 1961.

2.06 Peck, R., W. Hanson, T. H. Thornburn, Foundation Engineering, Wiley, New York, 1953, pp. 34-46.

2.07 Peech, M. et al., Methods of Soil Analysis for Soil-Fertility Investigations, U. S. Dept. of Agriculture, Circ. No. 757, 1947.

2.08 Report of Committee on Classification of Materials for Subgrades and Granular Type Roads, Proc. Highway Res. Board, Vol. 25, Wash., 1945, pp. 375-388. Discussion, pp. 388-392.

2.09 Road Research Laboratory, DSIR, Soil Mechanics for Road Engr., Her Majesty's Stationery Office, London, 1952, pp. 66-88.

2.10 Thorb, J., and Guy Smith, Soil Science, Vol. 67, No. 1, Jan. 1949, pp. 117-126.

2.11 U. S. Dept. of Agriculture, Mechanical Analysis of Soils, Bull. No. 4, Wash., 1896.

CHAPTER 3

Rheology

3.01 Fredrickson, A. G., Principles and Applications of Rheology, Prentice-Hall, Englewood Cliffs, N. J., 1964.

3.02 Freudenthal, A. M., The Inelastic Behavior of Engineering Materials and Structures, Wiley, New York, 1950.

3.03 Fung, Y. C., Foundations of Solid Mechanics, Prentice-Hall, Englewood Cliffs, N. J., 1965.

3.04 Green, A. E, and R. S. Rivlin, The Mechanics of Nonlinear Materials with Memory, Arch. Rat. Mech., No. 1, 1957, pp. 1–21.

3.05 Šuklje, L., Rheological Aspects of Soil Mechanics, Wiley-Interscience, New York, 1969, pp. 24–41.

3.06 Tuma, Jan J., Engineering Mathematics Handbook, McGraw-Hill, New York, 1970, p. 164.

3.07 Tuma, Jan J., Theory and Problems of Structural Analysis, McGraw-Hill, New York, 1969, pp. 138–144.

3.08 Yong, R. N. and B. P. Warkentin, Introduction to Soil Behavior, MacMillan, New York, 1966, pp. 80–94.

CHAPTER 4

Soil Water at Rest

4.01 Buckingham, E., Studies on Movement of Soil Moisture, U. S. Dept. of Agr., Bur. Soils, Bull. 38, 1907.

4.02 Capper, P. L., and W. F. Cassie, The Mechanics of Engineering Soils, E. F. Spon, London, 1964.

4.03 Capper, P. L., W. F. Cassie, and J. D. Geddes, Problems in Engineering Soils, E. F. Spon, London, 1966.

4.04 Cockrell, William D., Industrial Electronics Handbook, McGraw-Hill, New York, 1958, pp. 150–152.

4.05 International Critical Tables, Vol. 3, McGraw-Hill, New York, 1928, pp. 25–26.

4.06 Jumikis, A. R., Soil Mechanics, Van Nostrand, Princeton, N. J., 1962, pp. 192–208.

4.07 Peck, R. B., Walter Hanson, and T. H. Thornburn, Foundation Engineering, Wiley, New York, 1953, pp. 60–65.

4.08 Road Research Laboratory, D. S. I. R., Soil Mechanics for Road Engineers, Her Majesty's Stationery Office, London, 1952, pp. 293–325.

4.09 Russel, M. B., and M. G. Spangler, The Energy Concept of Soil Moisture and Mechanics of Unsaturated Flow, Proc. Highway Res. Board, Vol. 21, 1941, pp. 435–449.

4.10 Russell, M. B., The Utility of Energy Concept of Soil Moisture, Proc. Amer. Soil Sci., Vol. 7, 1942, pp. 90–94.

4.11 Schofield, R. K., The pF of Water in Soil, Trans. 3rd Int. Congr. Soil Sci., Vol. 2, Oxford, England, 1935, pp. 37–48.

CHAPTER 5

Soil Water in Motion

5.01 Capper, P. L., and W. F. Cassie, The Mechanics of Engineering Soils, E. F. Spon, London, 1964.

5.02 Capper, P. L., W. F. Cassie, and J. D. Geddes, Problems in Engineering Soils, E. F. Spon, London, 1966.

5.03 Casagrande, A., Seepage Through Dams; Contributions to Soil Mechanics, J. Boston Soc. Civil Engrs., July, 1937.

5.04 Casagrande, A., Seepage Through Dams, Journal of the New England Water Works Association, Vol. 1, No. 2, June 1937, p. 139.

5.05 Casagrande, A., and W. L. Shannon, Base Course Drainage for Airport Pavements; Transactions, ASCE, Vol. 117, 1952, pp. 792–816, and 1952a, p. 807.

5.06 Cedergren, H. R., Seepage, Drainage, and Flow Nets, Wiley, New York, 1967.

5.07 Conversion Factors for Engineers, Dorr-Oliver Inc., Stamford, Conn., 1961.

5.08 Darcy, H., Les Fontaines Publiques De La Ville De Dijon, Dalmont, Paris, 1856, pp. 674–8.

5.09 Forchheimer, P., Hydraulik, 3rd Ed. Teubner, Leipzig, 1930.

5.10 Gilboy, G., Hydraulic Fill Dams, Proceedings, Inter. Comm. Large Dams, Stockholm, 1933.

5.11 Harr, M. E., Groundwater and Seepage, McGraw-Hill, 1962.

5.12 Hazen, A., Physical Properties of Sands and Gravels with Reference to Their Use in Filtration, Report of State Board of Health, Boston, Mass., 1892.

5.13 Hough, K. B., Basic Soil Engineering, Ronald Press, New York, 1957, p. 69.

5.14 Hvorslev, M. J., Time Lag and Soil Permeability in Ground-Water Observations, Corps of Engrs., Waterways Expt. Station, Vicksburg, Miss., Bull. No. 36, 1951.

5.15 International Critical Tables, Vol. V, McGraw-Hill, New York, 1929, p. 10.

5.16 Jumikis, A. R., Soil Mechanics, Van Nostrand, Princeton, N. J., 1962, pp. 285–334.

5.17 Kozeny, J., Theorie und Berechnung der Brunnen; Wasserkraft and Wasserwirtschaft, 1933, Vol. 28, p. 104.

5.18 Lambe, T. W. and R. V. Whitman, Soil Mechanics, Wiley, New York, 1969.

5.19 Loudon, A. G., The Computation of Permeability from Simple Soil Tests Geotechnique, Vol. 3, London, 1952–53, pp. 165–182.

5.20 Scheidegger, A. E., The Physics of Flow Through Porous Media, MacMillan, New York, 1957.

5.21 Poiseuille, J. L., Recherches expérimentales sur le mouvement des liquides dans les tubes de très petits diamèteres, Comptes Rendus, Vol. 11, Paris, 1840, Also Memoires des Savants Etrangers, Vol. 9, Paris, 1846.

5.22 Singh, Alan, Soil Engineering in Theory and Practice, Asia Pub. House, London, 1967.

5.23 Taylor, D. W., Fundamentals of Soil Mechanics, Wiley, New York, 1948.

5.24 Terzaghi, K., Theoretical Soil Mechanics, Wiley, New York, 1943.

5.25 Terzaghi, K., and R. B. Peck, Soil Mechanics in Engineering Practice, Wiley, New York, 1943.

5.26 Wenzel, L. K., Methods for Determining Permeability of Water-Bearing Materials, U. S. Dept. of the Interior, Govt. Printing Office, Wash. D. C., 1942.

CHAPTER 6

Volume Change

6.01 Anagnosti, P., Stresses, Deformations, and Pore Water Pressure in Triaxial Tests Obtained by a Suitable Rheological Model, Proc. Eur. Conf. Soil Mech. Found. Engng., Wiesbaden, Vol. 1, 1963, pp. 1-6.

6.02 Casagrande, A., Classification and Identification of Soils, Proc. ASCE, Vol. 73, 1947, pp. 794-797.

6.03 Casagrande, A., The Determination of the Preconsolidation Load and its Practical Significance, Proc. First Intern. Conf. Soil Mech. and Found. Engr., Vol. 3, Harvard University Press, 1936, pp. 60-64.

6.04 Bowles, J., Engineering Properties of Soils and Their Measurement, McGraw-Hill, New York, 1970, pp. 97-102.

6.05 Gibson, R. E., and K. Y. Lo, A Theory of Consolidation for Soils Exhibiting Secondary Compression, Norwegian Geotechnical Institute, Oslo, Publication No. 41, 1961.

6.06 Peck, R. B, Walter Hanson, and T. H. Thornburn, Foundation Engineering, Wiley, New York, 1953, pp. 68-70, 265-280.

6.07 Schiffman, P. L., and R. E. Gibson, The Consolidation of Non-Homogeneous Clay Layers, Proc. ASCE, Journal of the Soil Mechanics and Foundation, Vol. 90, SM 5, 1964, pp. 1-30.

6.08 Schmertmann, J. H., Estimating the True Consolidation Behavior of Clay from Laboratory Test Results, Proc. ASCE, Vol. 79, No. 311, Oct. 1953; Discussion Vol. 81, No. 658, March, 1955.

6.09 Singh, Alan, Soil Engineering in Theory and Practice, Asia Pub. House, London, 1967, pp. 224-278.

6.10 Skempton, A. W., An Investigation of the Bearing Capacity of a Soft Clay Soil, J. Inst. of Civil Engrs., Vol. 18, London, 1942, pp. 307-309.

6.11 Skempton, A. W., Notes on the Compressibility of Clays, Quart. Jour. Geol. Soc., Vol. 100, London, 1944, pp. 119-135.

6.12 Skempton, A. W., and L. Bjerrum, A Contribution to the Settlement Analysis of Foundations on Clay, Geotechnique, Vol. 7, No. 4, London, 1957, pp. 168-178.

6.13 Šuklje, L., Rheological Aspects of Soil Mechanics, Wiley-Interscience, New York, 1969, pp. 151-183.

6.14 Taylor, R. W., Fundamentals of Soil Mechanics, Wiley, New York, 1948, pp. 208-251.

6.15 Taylor, D. W., and W. Merchant, A Theory of Clay Consolidation Accounting for Secondary Compression, J. Math and Physics, Vol. 19, 1940, pp. 167-185.

6.16 Terzaghi, K., Erdbaumechanik auf Boden-physicalischen Grundlagen, Deuticke, Vienna, 1925.

6.17 Terzaghi, K., Theoretical Soil Mechanics, Wiley, New York, 1963, pp. 265-296.

6.18 Terzaghi, K., and O. K. Fröhlich, Theory der Setzungen von Tonschichten, Deuticke, Leipzig, 1936, p. 166.

6.19 Terzaghi, K., and R. B. Peck, Soil Mechanics in Engineering Practice, Wiley, New York, 1948, pp. 173-182.

CHAPTER 7

Stress Analysis

7.01 Boussinesq, J., Application des potentiels à l'étude de l'équilibre et du movement des solides elastiques, Mem. Soc. Sci. Agric. Arts, Lille, 1885, p. 212.

7.02 Fadum, R. E., Influence Values for Estimating Stresses in Elastic Foundations, Proceedings Second International Conference on Soil Mechanics and Foundation Engineering, Vol. 3, Rotterdam, 1948.

7.03 Foster, C. R., and R. G. Ahlvin, Stresses and Deflections Induced by a Uniform Circular Load, Proceedings Highway Research Board, Vol. 34, 1954, p. 320.

7.04 Jumikis, A. R., Soil Mechanics, Van Nostrand, Princeton, N. J., 1962.

7.05 Newmark, N. M., Influence Charts for Computation of Stresses in Elastic Foundations, University of Illinois Eng. Experiment Station Bull. 338, 1942.

7.06 Timoshenko, S., and J. N. Goodier, Theory of Elasticity, McGraw-Hill, New York, 2nd ed., 1951, pp. 343-372.

CHAPTER 8

Strength Analysis

8.01 Bishop, A. W., Discussion on Measurement of Shear Strength of Soils, Proc. Conf. on the Measurement of Shear Strength of Soils, Geotechnique, Vol. 2, London, 1950, p. 113.

8.02 Bishop, A. W., and D. J. Henkel, The Measurement of Soil Properties in the Triaxial Test, Arnold, London, 1962.

8.03 Bjerrm, L., and N. E. Simons, Comparison of Shear Strength Characteristics of Normally Consolidated Clays, Proc. Conf. Shear Strength of Cohesive Soils, Amer. Soc. of Civil Engr., June 1960, pp. 711-726.

8.04 Bowles, Joseph, Engineering Properties of Soils and Their Measurement, McGraw-Hill, New York, 1970.

8.05 Lambe, T. W., and R. V. Whitman, Soil Mechanics, Wiley, New York, 1969.

8.06 Road Research Laboratory, Soil Mechanics for Road Engineers, Her Majesty's Stationery Office, London, 1956, p. 363.

8.07 Skempton, A. W., and A. W. Bishop, The Measurement of the Shear Strength of Soils, Geotechnique, Vol. 2, London, 1950-51, pp. 13-32.

8.08 Skempton, A. W., From Theory to Practice in Soil Mechanics, Wiley, New York, 1960, pp. 42-53.

8.09 Lambe, T. M., Soil Testing for Engineers, Wiley, New York, 1951.

8.10 Peck., R. B., Walter Hanson, and T. H. Thornburn, Foundation Engineering, Wiley, New York, 1953.

8.11 Reynolds, H. R., and P. Protopapadakis, Practical Problems in Soil Mechanics, Ungar, New York, 1959.

8.12 Singh, Alan, Soil Engineering In Theory and Practice, Asia Publishing House, London, 1967.

8.13 Smith, G. N., Elements of Soil Mechanics, Crosby Lockwood, London, 1968, pp. 73-107.

8.14 Terzaghi, K., Theoretical Soil Mechanics, Wiley, New York, 1943.

8.15 Tuma, Jan J., Engineering Mathematics Handbook, McGraw-Hill, New York, 1970, p. 57.

CHAPTER 9

Stability Analysis

9.01 Bjerrum, L., Progressive Failure in Slopes of Overconsolidated Plastic Clay and Clay Shales, Proc. ASCE, Vol. 93, No. SM 5 (Part 1), 1967, pp. 1-49.

9.02 Bishop, A. W., The Use of the Ship Circle in the Stability Analysis of Earth Slopes, Geotechnique, Vol. 5, 1955, pp. 7-17.

9.03 Bishop, A. W., and L. Bjerrum, The Relevance of the Triaxial Test to the Solution of Stability Problems, Proc., Amer. Soc. of Civil Engr. Research Conference on Shear Strength of Cohesive Soils, Boulder, Colorado, 1960, pp. 437-501.

9.04 Capper, P. L., W. F. Cassie, and J. D. Geddes, Problems in Engineering Soils, E. N. Spon, London, 1966.

9.05 Corps of Engineers, Engineering and Design of Rigid Pavements, SM 1110-45-303, 1958.

9.06 Culmann, K., Die Graphische Statik, Zurich, 1866.

9.07 Fellenius, W., Calculation of the Stability of Earth Dams, Transactions, 2d Congress on Large Dams, held in Washington, D. C., 1936, U. S. Govt. Printing Office, Washington, D. C., 1938, Vol. 4, pp. 445-462.

9.08 Hoffman, O, and G. Sachs, Introduction to the Theory of Plasticity for Engineers, McGraw-Hill, New York, 1953, p. 133.

9.09 Karol, R. H., Soils and Engineering, Prentice-Hall, N. J., 1960.

9.10 Lambe, T. W., and Robert V. Whitman, Soil Mechanics, Wiley, New York, 1969.

9.11 Meyerhof, G. G., Shallow Foundations, Proc. ASCE, Vol. 11, No. SM 2, 1965, pp. 21-31.

9.12 Meyerhof, G. G., The Ultimate Bearing Capacity of Foundations, Geotechnique, Vol. 2, London, 1951, pp. 301-332.

9.13 Morgenstern, N., Stability Charts for Earth Slopes During Rapid Drawdown, Geotechnique, Vol. 13, 1963, pp. 121-131.

9.14 Morgenstein, N. R. and V. E. Price, The Analysis of the Stability of General Slip Surfaces, Geotechnique, Vol. 15, 1965, pp. 79-93.

9.15 Singh, Alan, Soil Engineering in Theory and Practice, Asia Publishing House, London, 1967.

9.16 Peck, R. B., W. E. Hanson, and T. H. Thornburn, Foundation Engineering, Wiley, New York, 1953.

9.17 Petterson, K. E., The Early History of Circular Sliding Surfaces, Geotechnique, London, Vol. 5, 1955, pp. 275-296.

9.18 Prandtl, L., Über Die Eindrigung Festigkeit (Harte) Plastischer Baustoffe und Die Festigkeit Von Schneiden, Zeitschrift Für Angewandte Mathematik and Mechanik, Vol. 1, 1921, pp. 15-20.

9.19 Reynolds, H. R., and P. Protopapadakis, Practical Problems in Soil Mechanics, Ungar, New York, 1959.

9.20 Road Research Laboratory, Soil Mechanics for Road Engineers, Her Majesty's Stationery Office, London, 1952, pp. 376-390.

9.21 Smith, G. N., Elements of Soil Mechanics, Crosby Lockwood, London, 1968.

9.22 Sokolovski, V. V., Statics of Soil Media, Butterworths, London, 1960.

9.23 Taylor, D. W., Stability of Earth Slopes, J. Boston Soc. of Civil Engrs., Vol. 24, p. 197.

9.24 Taylor, D. W., Fundamentals of Soil Mechanics, Wiley, New York, 1948, pp. 406-476, 572-584.

9.25 Terzaghi, K., Evaluation of Coefficients of Subgrade Reaction, Geotechnique, Vol. 5, No. 4, London, Dec. 1955, pp. 297-326.

9.26 Terzaghi, K., Stability of Slopes of Natural Clay, Proc. Int. Conf. Soil Mechanics, Vol. I, Harvard University, 1936.

9.27 Terzaghi, K., Theoretical Soil Mechanics, Wiley, New York, 1954, pp. 118-136.

CHAPTER 10

Frost Action

10.01 Aldrich, Harl P., Frost Penetration Below Highway and Airfield Pavements, Highway Research Board, Vol. 36, Bul. No. 135, 1956, pp. 124-144.

10.02 Berggren, W. P., Prediction of Temperature Distribution in Frozen Soil, Transactions, American Geophysical Union, Part III, 1943.

10.03 Carlson, H., and M. S. Kersten, Calculation of Depths of Freezing and Thawing Under Pavements, Highway Res. Board, Vol. 33, Bull. No. 71, 1953.

10.04 Corps of Engineers, Engineering and Design, Pavement Design for Frost Conditions, EM-1110-345-306, 1951.

10.05 Highway Research Board, Frost Action in Soils, A Symposium, Highway Research Board, Washington, D. C., Special Report No. 2, 1952.

10.06 Kersten, M. S., Final Report, Laboratory Research for the Determination of the Thermal Properties of Soils, Engr. Expt. Station, University of Minnesota, 1949.

10.07 Penner, E., The Mechanism of Frost Heaving in Soils, Highway Research Board, Vol. 39, Bull. No. 225, 1959, pp. 1-22.

10.08   Road Research Laboratory, Soil Mechanics for Road Engineers, Her Majesty's Stationery Office, London, 1952.

10.09   Tystovich, N. A., Bases and Foundations on Frozen Soils, Highway Research Board, Vol. 38, Special Report No. 58, 1960.

10.10   Yong, R. N., and B. P. Warkentin, Introduction to Soil Behavior, MacMillan, 1966, pp. 392–427.

10.11   Yoder, E. J., Principles of Pavement Design, Wiley, New York, 1959.

glossary
of
symbols

$a, b, c, \ldots$	segments (cm)
a	diffusion constant (cm²/sec)
b	width (cm)
c	cohesion (gm/cm²)
d	diameter (cm)
e	base of natural logarithm (dimensionless)
e	void ratio (dimensionless)
h	capillary rise (cm)
h	hydraulic head (cm)
i	hydraulic gradient (dimensionless)
i	Newmark's influence value (dimensionless)
i	thermal gradient (dimensionless)
k	permeability (cm/sec)
k	spring constant (gm/cm²)
k	thermal conductivity (Btu/Hr/cm/°F)
l	length (cm)
m	shape parameter (dimensionless)
n	porosity (dimensionless)
n	shape parameter (dimensionless)
p	intensity of pressure (gm/cm²)
p	critical normal stress (gm/cm²)

q	intensity of distributed load (gm/cm²)
q	intensity of fluid discharge (cm/sec)
q	intensity of heat discharge (Btu/Hr/cm²)
r	radius (cm)
s	critical shearing stress (gm/cm²)
t	time (sec)
u	excess hydrostatic pressure (gm/cm²)
v	discharge velocity (cm/sec)
w	water content (dimensionless)
A	area (cm²)
C	constant (any dimension)
D	diameter of soil particle (cm)
E	modulus of elasticity (gm/cm²)
F	freezing index (degree-days)
G	specific weight (dimensionless)
H	hydraulic head (cm)
H	thickness of soil layer (cm)
K	Boussinesq's stress factor (dimensionless)
L	latent heat of fusion (Btu/cm³)
L	length (cm)
P	concentrated load (gm)
Q	fluid discharge (cm³/sec)
Q	heat discharge (Btu/Hr)
R	radius (cm)
S	degree of saturation (dimensionless)
T	temperature (degrees)
T	resultant working force (gm)
U	average value of consolidation ratio (dimensionless)
V	volume (cm³)
W	total weight (gm)
α, β, γ	angles (degrees)
γ	unit weight (gm/cm³)
δ	linear displacement (cm)
ϵ	strain (dimensionless)
η	dashpot constant (gm/cm² × sec)
θ	angle of slip plane (degrees)
μ	Poisson's ratio (dimensionless)
ν	viscosity (gm/cm² × sec)
σ	stress (gm/cm²)
ϕ	angle of friction (degrees)
ϕ	exponent of turbulence (dimensionless)
ϕ	stress influence value (dimensionless)

COMPOSITE SYMBOLS

a_v	coefficient of compressibility (cm²/gm)
c_v	coefficient of consolidation (cm²/sec)
e_0	initial void ratio (dimensionless)
e_c	critical void ratio (dimensionless)
h_c	capillary rise (cm)
i_c	critical hydraulic gradient
p_0	initial pressure (gm/cm²)
p_c	critical pressure (gm/cm²)
q_u	ultimate bearing capacity (gm/cm²)
$r_{ij}, \omega_{ij}, z_{ij}$	polar coordinates (cm, deg., cm)
u_w	pore-water pressure (gm/cm²)
v_a	approach velocity (cm/sec)
v_f	seepage velocity (cm/sec)
x_{ij}, y_{ij}, z_{ij}	cartesian coordinates (cm)
z_p	frost-penetration depth (cm)
C_c	compression index (dimensionless)
C_c'	approximate compression index (dimensionless)
C_g	coefficient of gradation (dimensionless)
C_s	swelling index (dimensionless)
C_u	coefficient of uniformity (dimensionless)
E_0	modulus of subgrade (gm/cm²)
F_s	safety factor (dimensionless)
G_m	specific gravity of soil sample (dimensionless)
G_s	specific gravity of solids (dimensionless)
G_w	specific gravity of water (dimensionless)
N_c, N_γ, N_q	bearing capacity factors (dimensionless)
N_ϕ	flow value (dimensionless)
S_{ji}	stress influence factor (dimensionless)
T_s	surface tension (gm/cm)
T_v	consolidation time factor (dimensionless)
U_z	consolidation ratio (dimensionless)
V_g	volume of gases (air) (cm³)
V_m	volume of soil sample (cm³)
V_s	volume of solids (cm³)
V_v	volume of voids (cm³)
V_w	volume of water (cm³)
W_g	weight of gases (air) (gm)
W_m	weight of soil sample (gm)
W_s	weight of solids (gm)
W_w	weight of water (gm)

γ_d	dry unit weight (gm/cm^3)
γ_m	unit weight of soil sample (gm/cm^3)
γ_0	unit weight of distilled water at 4°C (gm/cm^3)
γ_s	unit weight of solids (gm/cm^3)
γ_{sat}	unit weight of saturated soil sample (gm/cm^3)
γ_w	unit weight of water (gm/cm^3)
ϵ_0	initial strain (dimensionless)
ϵ_E	elastic strain (dimensionless)
ϵ_Y	yield strain (dimensionless)
ν_w	viscosity of water (gm/cm$^2 \times$ sec)
ρ_z	shape function, circular load (dimensionless)
$\sigma_r, \sigma_t, \sigma_z$	normal stresses in polar system (gm/cm^2)
$\sigma_{rt}, \sigma_{tz}, \sigma_{zr}$	shearing stresses in polar system (gm/cm^2)
$\sigma_x, \sigma_y, \sigma_z$	normal stresses in cartesian system (gm/cm^2)
$\sigma_{xy}, \sigma_{yz}, \sigma_{zy}$	shearing stresses in cartesian system (gm/cm^2)
$\sigma_1^p, \sigma_2^p, \sigma_3^p$	principal normal stresses (gm/cm^2)
σ_E	elastic stress (gm/cm^2)
σ_0	initial stress (gm/cm^2)
σ_y	yield stress (gm/cm^2)
$\bar{\sigma}_{zz} = p$	critical normal stress (gm/cm^2)
$\bar{\sigma}_{zx} = s$	critical shearing stress (gm/cm^2)
τ_{ji}, τ_{ij}	time interval ji, ij (sec)
$\psi_z(m,m)$	shape function, line load (dimensionless)
$\phi_z(m,n)$	shape function, rectangular load (dimensionless)

COMBINED SYMBOLS

GI	group index (dimensionless)
LI	liquidity index (dimensionless)
LL	liquid limit (dimensionless)
PI	plasticity index (dimensionless)
PL	plastic limit (dimensionless)
SI	shrinkage index (dimensionless)
SL	shrinkage limit (dimensionless)

MATRIX SYMBOLS

$\hat{T}_{ji}, \hat{T}_{ij}$	state vector transport matrices in the interval τ_{ji}, τ_{ij}
T^m, T^n	stress tensors in m, n-system
$\hat{Y}_j, \hat{Y}_i$	state vectors at stations j, i
Π^{mn}, Π^{nm}	angular transformation matrices mn, nm

author
index

AHLVIN, R. G., *182, 318*
ALDRICH, H. P., *291, 294, 295, 320*
ALLEN, H., *314*
ANAGNOSTI, P., *146, 317*
ATTERBERG, A., *8*

BERGGREN, W. P., *292, 294, 320*
BISHOP, A. W., *318, 319*
BJERRUM, L., *318, 319*
BOUSSINESQ, J., *171, 187, 318*
BOWLES, J. E., *148, 317, 318*
BUCKINGHAM, E., *78, 315*

CAPPER, P. L., *266, 273, 315, 316, 319*
CARLSON, H., *320*
CASAGRANDE, A., *17, 32, 33, 101, 103, 117, 313, 314, 316, 317*
CASSIE, W. F., *266, 273, 315, 316, 319*
CEDERGREN, H. R., *316*
COCKRELL, D., *78, 315*
CULMANN, K., *249, 250, 319*

DARCY, H., *100, 103, 316*
DERESIEWICZ, H., *19, 313*
DUNHAM, C. W., *313*
DUPUIT, J., *108*

327

subject index